Inclusive University Built Environments

Itab Shuayb

Inclusive University Built Environments

The Impact of Approved Document M
for Architects, Designers, and Engineers

 Springer

Itab Shuayb
American University of Beirut
Beirut, Lebanon

ISBN 978-3-030-35860-0   ISBN 978-3-030-35861-7   (eBook)
https://doi.org/10.1007/978-3-030-35861-7

This Springer imprint is published by the registered company Springer Nature Switzerland AG
The registered company address is: Gewerbestrasse 11, 6330 Cham, Switzerland

This book is dedicated to Ms. Aya Aghabi, the founder of Accessible Jordan, who passed away a few months ago. We shared the mission of making the built environment inclusive and accessible for all. Despite the tragic news of Aya's passing, her soul remains at the core of the social demand to continue her legacy, mission and global movement for promoting inclusive ethos in our built environment. The book commemorates Aya's legacy by adopting the inclusive design approach and proposing alternative design solutions to tackle accessibility barriers that affect a wide range of users, including individuals with disabilities at universities.

Preface

This book arose from my personal interest in inclusive design and barriers that can restrict users from integrating in their societies. Inclusive design became a personal interest from 1993, when my sister had a car accident and became a wheelchair user. From then onwards, I realised how architecture can disable users and restrict them from gaining access to education, employment and services. Since April 1993, my sister could no longer join her classmates at her school. Schools and universities in Lebanon were inaccessible for disabled people and thus my sister had to study at home to pursue her school and university education. In 2001, my sister was admitted to the MA programme in counselling at the University of Newcastle-upon-Tyne in the UK. This year was a remarkable year, as it was the first time after seven years of home schooling, that my sister was treated on an equal basis with her classmates, and although the Department of Counselling was not completely accessible, the faculty members used to relocate their classrooms into accessible ones to comply with their legal duties laid down by the British Disability Discrimination Act (DDA) 1996. Recognising that architectural features of an environment can be a barrier for end users, such as mothers and children, older people and individuals with disabilities, gave me the incentive to write about the impact of architectural design standards and Building Regulations on the built environment at universities and ways to enhance accessibility for all users.

This book focuses on examining accessibility in the educational sector in the UK to investigate whether adopting an inclusive design approach in a university setting is preferable to just meeting legal building requirements. Six building case studies at the University of Kent were selected in order to investigate whether the design solutions had addressed the needs of a wide range of users. Moreover, the book investigates the impact of the legislation and Building Regulations on six different university buildings dating from six different decades, the 1960s, 1970s, 1980s, 1990s, 2000s and 2010s, at the Universities of Essex, Bath and Kent to determine whether they have achieved inclusive design .The book then sets out a proposal to deliver the benefits of adopting the inclusive design approach by recommending alternative design solutions to tackle accessibility barriers that affect a wide range of users, including individuals with disabilities at the University of Kent.

The first part of the book explores the identification and assessment methods used to investigate the level of accessibility at six main buildings at the University of Kent. Such methods included carrying out access audits on six buildings, personal interviews with ten individuals with disabilities, four education providers and two architects, and conducting online surveys. The second part deals with the feedback on the first phase results and the consultations with individuals with disabilities. It also investigates the main accessibility barriers at the four buildings identified by users who were interviewed as the most problematic with regard to segregation and exclusion. Case studies of two buildings at Kent University were selected by users who were interviewed for redesign by the author with the aim of achieving inclusion.

Data analysis of the two phases of the research showed that the built environment at the University of Kent did not cater for all users. Observations from access audits and feedback from consultations with users, including those with disabilities, suggested that the physical environment on campus and inside buildings needed to be changed to cater for sensory and cognitive impairments. Moreover, feedback from consultation with individuals with disabilities and education providers highlighted the importance of reviewing the management practices and procedures with staff members from across departments and buildings, so as to improve facilities and services and thus attract more disabled applicants. Insufficient training in disability awareness was one of the key reasons that staff members and architects failed to cater for the specific needs of disabled people. A key finding was that the inclusive design approach at universities is seen as preferable by students, staff and other stakeholders to just satisfying the regulations relating to accessibility.

These results highlight the market demand for universities to base their entire business on the inclusive design strategy. By recognising the diversity of their users' life styles and learning about their individual experiences, universities can identify and remove the physical and mental barriers to achieving an inclusive university environment.

Beirut, Lebanon   Itab Shuayb

Contents

Chapter 1
Introduction to Inclusive University Built Environments

1.1 Introduction

For many decades, since the introduction of a range of antidiscrimination disability legislation, the design of the built environment in the UK has been influenced by the needs of people with disabilities. The British Disability Discrimination Act 1995, its amendment in 2005 and the Equality Act 2010 have caused a shift in attitudes towards creating built environments that are accessible for people with disabilities (Goldsmith 1997; Imrie and Hall 2001; Steinfeld and Maisel 2012).

It was over 16 years after introducing the Chronically Sick and Disabled Persons Act in Britain in 1970 that design guidance was introduced in order to implement the law and eliminate architectural barriers for people with disabilities (Goldsmith 1997, pp. 101, 121).

The British Approved Document Part M was introduced to address issues of access to newly constructed facilities so they become accessible to, and usable by, people with disabilities. These guidelines and regulations were revised to cover a range of facilities to meet the needs of disabled people in the public and private sectors and were useful tools for designers, enabling them to determine appropriate design parameters or space requirements and essential design features. However disability acts aimed at eliminating discrimination against individuals with disability have not always been accompanied by building regulations that ensure inclusion.

The process of designing accessible environments for individuals with disabilities has historically passed through different phases, which are often described as models. The two most relevant ones in this book are the medical and the social models, which will be examined in depth in Chap. 2. These models have shaped the approach to the design of the built environment.

Prior to the disability rights movement that started in the 1960s, disability in Britain centred on the medical model, which restricted many individuals with disabilities from interacting with the society and the environment, since their impairment was considered to be the main factor preventing interaction.

© Springer Nature Switzerland AG 2020
I. Shuayb, *Inclusive University Built Environments*,
https://doi.org/10.1007/978-3-030-35861-7_1

In the absence of a medical cure for their physical condition, individuals with disabilities had to attend welfare facilities, such as special rehabilitation centres, special schools and welfare units, which resulted in exclusion and segregation from the mainstream society.

According to Simon Jarret (2012, p. 28), in Britain, the Industrial Revolution led to mass migration to cities and social pressures, which resulted in the creation of the Poor Laws in 1834 and the building of 'asylums'-purpose built institutions to house people with mental health difficulties, who were described at that time as 'idiots' or 'lunatics'. The development of segregated institutions continued into the early twentieth century, although the purpose of moving people to institutions changed. During the period between the two World Wars, laws were passed to promote segregation of people with disabilities, especially those with learning disabilities, from the mainstream society. These laws encouraged the building of special schools for children with learning disabilities, which promoted segregation and isolation. In the 1930s, the IQ test was introduced and people scoring low marks in the test were categorised as 'mentally defective' and were prohibited from attending even special schools.

The introduction of the British National Health Service in 1946 and defining of disability according to the medical model led to the introduction of a new term, 'mentally handicapped', which resulted in the replacement of special institutions and schools with hospitals.

Attitudes towards people with mental health difficulties and physical disabilities had shifted from seeing them as 'dangerous or degenerate' to viewing them more sympathetically as people in need of treatment, although still a drain on the public purse. People with a learning disability remained segregated and isolated, and the standard of care was extremely poor. In 1920, the Education Act was introduced to promote special education for children with disabilities. The Act mandated that local authorities should be responsible for providing special education for blind and deaf children (Jarret 2012, p. 37). This situation prompted many individuals with disabilities to call for equal rights which emphasised the right of choice and of opportunity to have access to education, employment, transport and public and private services. Acknowledging that the special institutions were a major barrier to the inclusion of individuals with disabilities in the mainstream society, the disability rights movement, which started in the USA, has led to a shift in attitudes towards promoting inclusion and gaining access to education, employment and services.

This movement led advocates of accessibility and architects in the USA to work on eliminating physical barriers in 1956, and the University of Illinois Champaign-Urbana was the first university in the world to attempt to enhance accessibility to its buildings for disabled people. Moreover, the independent living movement, which started in the USA at the University of California Berkeley in 1972, called for the removal of architectural and transportation barriers that prevent people with disabilities from having equal opportunities to share fully in all aspects of society. Hundreds of independent-living centres and units were established across the USA, and throughout much of the rest of the world (Goldsmith 1997, pp. 54–57). This movement, along with disability rights and access advocacy, called for the need to design buildings accessible to people with disabilities and raised awareness of the need to bring people with disabilities into the mainstream society, ensuring equal opportunity and eliminating barriers to access to, and use of, the built environment (Steinfeld and Maisel 2012, p. 15).

The social model of disability changed the way of thinking about disability as it shifted the issue from the individual and placed it on society. Instead of disability being the bad luck or fault of the disabled person, disability has become the fault of society which does not provide an enabling environment, resulting in the disempowerment of certain groups in society (Finkelstein 1981; Barnes 1991; Oliver 1990, 1996); hence, the solution lies in the hands of society. One important component of this solution is the adoption of an inclusive approach to architectural design to eliminate barriers for users from different age groups and with different abilities (Goldsmith 1997; Imrie and Hall 2001).

As a result, advocates of accessibility began to put their efforts into enhancing the built environment by providing designs and environments that all people can use and enjoy, regardless of age group, abilities and capabilities. The new paradigm, known as 'inclusive design', goes beyond the necessity to create accessible environments for individuals with disabilities. It aims at creating environments that maintain quality of life and independent living for all potential users by eliminating barriers and avoiding stigma. To achieve this goal, inclusive design demands a shift in attitudes and perspective towards placing end users at the heart of the design process, acknowledging diversity and differences by offering choices and designs that cater for a wide range of users.

In spite of the major strides achieved in this field, many facilities and built environments still have significant barriers that restrict access for many users, such as children, mothers/carers with children and old people. Acknowledging that British legislation places duties on schools and universities to promote inclusion and eliminate adverse physical features, in compliance with the Approved Document M, this book investigates the impact of Approved Document M on different university buildings in the UK. A selection of six buildings at the Universities of Kent, Bath and Essex built during the six decades of the 1960s, 1970s, 1980s, 1990s, 2000s and 2010s are studied to investigate the impact of Approved Document M on changing building designs to ensure that buildings are accessible for all potential users, including people with disabilities. Moreover, the book investigates the challenges and responsibilities that restrict architects and education providers wishing to achieve inclusion. It also presents the University of Kent as a case study and sets out a proposal to deliver the benefits of the inclusive design approach by proposing adoption of alternative design solutions that tackle accessibility barriers at the University of Kent, which affect a variety of users, including individuals with disabilities.

The book is divided into eight chapters which present case studies of existing educational buildings and provide recommendations for achieving inclusive design in university environments.

This introductory chapter provides the background to the antidiscrimination disability legislation and its influence on the needs of disabled people. Consequently it discusses the medical and social models that have influenced disability and reveals the effects of these models on the built environment in educational settings.

Chapter 2 reviews the definitions of 'disability' and their relationship with the most recent and broad definitions of disability adopted in the UK. Next, the classification and characteristics of disability are discussed and reviewed. The chapter goes on to focus on the history of the human rights and disability rights movement in Britain.

Chapter 3 introduces the emergence of the design standard codes and their development. Then it defines inclusive design, its principles and importance, and goes on to present four case studies that have adopted the inclusive design approach to highlight its importance in promoting inclusion and integration for all potential users, including individuals with disabilities.

Chapter 4 addresses the research questions and methodological framework and design of the University of Kent case study. To investigate the impact of Approved Document M on university buildings, this chapter presents both the qualitative and quantitative data collection methods used. These two methods address critical issues surrounding the process of identifying and proposing inclusive designs and recommendations to accommodate the needs of all end users. The first phase included carrying out access audits on six buildings, personal interviews with ten individuals with disabilities, four education providers and two architects, and online surveys at the University of Kent. The second phase comprised feeding back Phase 1 results to the individuals with disabilities and analysing artefacts and architectural drawings of buildings.

Chapter 5 explores the identification and assessment procedures used at the University of Kent to investigate the level of accessibility at the selected buildings. It focuses on end users' contributions with regard to addressing the main accessibility barriers encountered. A further investigation discussed in this chapter focuses on the disability awareness of education providers and architects and how this is reflected in the provision of services that meet the legislative duties to eliminate physical barriers and promote inclusion among students and staff members.

Chapter 6 provides strategies for designing inclusive environments with respect to two different university buildings. Two different case studies, all centred on the inclusive design approach, offer the reader new insights into the way people interact with the built environment. The two buildings were selected with respect to building type, period of original construction, use, and historical and organisational types.

Chapter 7 presents an extensive discussion of the five main issues regarding accessibility to the universities which are the subject of the case studies, namely, (1) sociocultural differences and inclusive design, (2) misinterpretation of inclusive design and disability, (3) accessible design and regulation barriers, (4) procedural barriers and (5) organisational barriers.

Chapter 8 presents general recommendations and guidelines stemming from the methods used. It aims to explain how inclusive principles can be put into practice to achieve a friendly and welcoming university environment. This is followed by research limitations and research contributions to ways in which design of universities can support the creation of inclusive and sustainable environments.

This book is based on three main approaches. First is the adoption of the social model of disability to enhance accessibility and change attitudes towards a user-centred approach. Second is the inclusion of valuable information about the end users' experience of accessing the built environment and their input in proposing design solutions. The third approach in this book offers different design proposals that accommodate people's different abilities and needs to highlight the importance and value of adopting the inclusive design approach at universities with respect to the diversity of users.

The university building case studies presented in this book revealed that the universities did not fully embrace the inclusive design principles in tackling accessibility barriers. The findings highlight the fact that individuals with disabilities are socially and culturally marginalised and segregated when using university services and this produces bias in the built environment. Such bias is reflected in special provisions that reflect the dominance of the medical model of disability. Interviews with architects revealed that many of them preserve social and cultural attitudes in responding to the design needs of individuals with disabilities by relating impairments to medical conditions that are mainly concerned with mobility deficiency and impairments. The cultural bias towards responding mainly to wheelchair users' needs, rather than acknowledging the wide spectrum of disabilities, is reflected in architects' understanding of disability. Moreover, access audits revealed that the age of a building does not always correlate with its accessibility, since accessibility is only defined and understood by professionals and architects in terms of catering for one type of disability, namely physical disability or impaired mobility.

Interviews with individuals with disabilities and online surveys at the University of Kent highlighted the fact that physical and management barriers coexisted. There was an absence of disability knowledge and understanding of users' needs, and a failure to involve or consult users when a building was refurbished or renovated or new services were introduced. Many individuals with disabilities believed that eliminating physical barriers can solve part of the accessibility issue, but this is not sufficient to provide a completely accessible environment. A key element in achieving this, as many suggested, lies in providing appropriate and effective management and maintenance procedures that are monitored and checked regularly.

The interviews with individuals with disabilities, education providers and architects at the University of Kent highlighted the importance of education and training in raising awareness about types of disability and specific needs. Such awareness is vital in order to enhance services so that all users can benefit from them.

The recommendations of these stakeholders informed the redesigning of a number of buildings selected from the universities to achieve an inclusive environment. On the other hand, whilst architects believed that legislation and building regulations assisted them in eliminating architectural barriers, such regulations were not sufficient to achieve an inclusive built environment as they only anticipated the needs of people with physical and sensory impairments and did not take account of individuals' differences and diversities. Such an important finding calls for a different approach to design, namely a shift towards user-centred needs, which can lead to achieving a sustainable and inclusive environment that can benefit all users, regardless of their abilities and differences.

Ultimately, this book aims to play a part in bringing about a shift in attitudes towards the design of inclusive environments. The new century demands that architects and designers who are shaping university environments should shift from focusing only on accessibility provision towards improving functions for a broad range of people to promote inclusiveness and diversity. Architects oriented towards a user-centred approach can ensure that spaces and environments are used in such a way that the need for specific accommodations is reduced and any stigma is avoided by placing people with disabilities and able population on an equal playing footing.

Further research could be conducted on a larger number of universities with different topographical sites, building types and period of original construction. This book can provide guidance for architects and professionals, helping them to measure and record the design characteristics of a university in order to propose a more precise and inclusive university environment. This study provides recommendation for how to achieve an inclusive university centred on general aspects of enhancing accessibility to the campus, such as means of transportation, pedestrian routes, car parking spaces, internal facilities and emergency exit routes. However, it did not cover the neighbourhood surroundings, such as streets, public spaces, schools, leisure attractions and other properties. Further research could examine how these aspects can interact to achieve an inclusive environment. Moreover, it could focus on outdoor environments, such as roads, pavements, outdoor spaces and gardens across campuses, to determine how users interact with the outdoor environment and how such interaction can promote inclusiveness. This research is required to enhance the users' interaction with their built environment by creating inclusive facilities and environments that brings benefits to all.

References

Barnes, C. (1991). *Disabled people in Britain and discrimination*. London: Hurst and Co.

Finkelstein, V. (1981). To deny or not to deny disability. In A. Brechin et al. (Eds.), *Handicap in a social world*. Hodder and Stoughton: Sevenoaks.

Goldsmith, S. (1997). *Designing for the disabled, the new paradigm*. Oxford: Architectural Press.

Imrie, R., & Hall, P. (2001). *Inclusive design: Designing and developing accessible environments*. London: Spon Press.

Jarret, S. (2012). Disability in time and place. *English Heritage disability history web content*. Retrieved July 30, 2018, from https://content.historicengland.org.uk/content/docs/research/disability-in-time-and-place.pdf

Oliver, M. (1990). *The politics of disablement*. Basingstoke: Macmillan.

Oliver, M. (1996). *Understanding disability: From theory to practice*. Basingstoke: Macmillan.

Steinfeld, E., & Maisel, J. (2012). *Universal design: Creating inclusive environments*. Hoboken, NJ: John Wiley & Sons.

Chapter 2
Impact of the Disability Rights Movement and Legislation on Educational Programmes and Buildings

2.1 Introduction

The chapter reviews the definitions of 'disability' and its association with the most recent and broad definitions of disability adopted in the UK. Next, the classification and characteristics of disability are discussed and reviewed. Then this chapter goes on to focus on the history of human rights and the disability rights movement in Britain.

2.2 Definition of Disability

2.2.1 Models of Disability

The definition of the term 'disability' is complex and controversial. Various disability models and definitions have been put forward by disability activists and policy-makers. Smart and Smart (2006) have identified four different models of disability. The biomedical model defines disability in medical terms, indicating that it is an individual's own problem, which could be resolved by having medical treatment that would enable that person to integrate into society and gain access to services and facilities. The functional and environmental models of disability, as defined by Smart and Smart (2006), are interactional models in which the individual's disability and environment are considered to be the main disabling factors that prevent the person from interacting with the mainstream society. Thus disability is defined 'in relation to the skills, abilities and achievements of the individual in addition to biological and organic factors' (Smart and Smart 2006, p. 32), without taking into account any social and economic barriers which could restrict the person's ability to gain access to education, employment and facilities and services. The fourth model

© Springer Nature Switzerland AG 2020
I. Shuayb, *Inclusive University Built Environments*,
https://doi.org/10.1007/978-3-030-35861-7_2

of disability that Smart and Smart (2006) acknowledge is a 'sociopolitical model' that defines disability as:

> A social construction in that the limitations and disadvantages experienced by disabled people have nothing to do with their disability but are only related to social constructions and negative attitudes towards disabled people (Smart and Smart 2006, p. 34).

To eliminate the social stigma, prejudice and discrimination against and inferiority of disabled people, 'the sociopolitical model' asserts disabled people's rights to 'self-definition and self-determination, refusing to allow "medical experts" or "professionals" to define their disability to determine the outcomes or judge the quality of their lives' (Smart and Smart 2006, p. 34). Thus 'disability' in the sociopolitical model is viewed as a public concern rather than a personal tragedy.

On the other hand, Kaplan (1998), the Executive Director of the World Institute on Disability, proposes four models of disability which are related to the Smart and Smart (2006) models. Kaplan's first model, known as 'the moral model of disability', views disability as the outcome of 'a sin' in which disability is always associated with feelings of guilt and shame on the part of the entire relatives of whom the person with disability is a member. This model has resulted in segregating and isolating disabled people preventing them from integrating with their societies and prohibiting them from gaining access to schools, employment and public services. Swain and French (2000, p. 572) point out that 'the moral model of disability, or what they call the 'tragedy model', is dominant and infused throughout media representations, language, cultural beliefs, research, policy and professional practice'. The tragedy model of disability perceives disability as a personal problem related to impairment, rather than the failure of society to meet the person's needs by providing accessible facilities and services. Swain and French's (2000) 'tragedy model' assumes that disabled people cannot function properly in society unless they become 'normal' and their impairments are cured, although this assumption is rarely articulated by the people with disabilities themselves 'who acknowledge disability as a major part of their identity' (Swain and French 2000, p. 573).

The second model of disability that Kaplan (1998) defines is the 'medical model' which regards disability 'as a defect or sickness that has to be cured through medical intervention'. This model is similar to the 'biomedical model' of Smart and Smart (2006) as it relates disability to medical conditions that prevent the person from interacting with the facilities and services provided.

The 'rehabilitation model' is Kaplan's third model, which is a result of 'the medical model', since it regards disability as a deficiency that must be fixed by rehabilitation or other professional help (Pfeiffer 1998). This model was espoused after World War II as a result of the increased number of disabled veterans who needed rehabilitation so as to be able to participate in society again.

According to Kaplan (1998), the 'disability model' is the fourth model of disability. It defines disability as a normal aspect of life, which has been created by society and its indifference towards disabled people. This model matches the 'social model' of disability, which was developed in the 1970s by British activists in the Union of the Physically Impaired Against Segregation (UPIAS). 'The social model' proposed by (UPIAS) is based on the belief that people with disabilities are excluded and segregated from their communities as a result of the negative mindset of society which considers people with disabilities to be unable to integrate with society due to their 'impairment'. Such negative attitude has restricted many people with disabilities from enrolling at educational institutions and reduced their chances of attaining proper jobs and developing their community. Accordingly, the (UPIAS) 'social model of disability' acknowledges the fact that people with disabilities have impairment, which can limit their ability to move, see, hear or talk, but it does not prohibit them from gaining access to education, employment and public services, if their particular needs are considered when designing the built environment. For example, a child with low vision or visual impairment can gain access to school, if the school provides Braille or audio recording books. Similarly, a child with mobility impairment can attend school if the school is properly designed to provide level access, furniture and toilet that is accessible for a wheelchair user or a child using a walking aid.

A number of authors (Finkelstein 1980, 1981; Barnes 1991; Oliver 1990b, 1996) assert that disablement lies in the organisational and social polices of societies that provide services and facilities which do not anticipate the needs of disabled people. By creating built environments that do not take into account their physical impairments, disabled people are socially excluded and segregated from using facilities and services, resulting in a form of social oppression.

Riggar and Maki (2004) point out that the definitions of disability are important when it comes to defining eligibility for programmes and services. 'Defining disability is vital when attempting to influence policies that respond to different disability needs. Each of the models reviewed above reduces the concept of disability to a single value, which does not capture, explain or describe the different experiences of disabilities' (Riggar and Maki 2004, p. 26). Thus this review focuses on two main definitions or models which have been adopted in recent decades and which have shaped to a great extent the current approach to the built environment.

The first is the 'medical model of disability', which was adopted by the World Health Organization (WHO) in the 1980s when disability was defined in medical terms and was categorised according to individual experience, such as that of the blind, quadriplegic or mentally ill, without taking into consideration social justice. According to the 'medical model', disability is defined in terms of 'impairment', 'disability' and 'handicap' (Holmes-Siedle 1996). 'Impairment' as defined by the World Health Organization (WHO 1980) is 'a loss of, or abnormality in, structure or function, while "disability" is the inability to perform an everyday activity due to the specific impairment'. On the other hand, the term 'handicap', as defined by the World Health Organization, describes 'a person who is incapable of carrying out normal social activities because of an impairment or disability' (WHO 1980).

The 'medical model', or what Smart and Smart (2006) name the 'biomedical model', legitimises prejudice and discrimination against people with disabilities, and considers disability as a 'problem' or 'bad luck' for those individuals with disabilities. A good example of how 'the medical or the biomedical model' discriminates against individuals with disability is illustrated in this quotation.

When individuals with disabilities are not integrated in the workplace, schools, and other social institutions, their absence is usually not noticed. After all, according to this attribution theory, individuals with disabilities are thought to be responsible for their stigmatization. Clinicians have attempted to include environmental issues in their classification/diagnostic systems; however, the degree of prejudice and discrimination experienced or the lack of accommodations is typically not considered when medical professionals determine the level of severity of the disability or render a percentage of impairment (Smart 2005a, p. 3).

Moreover, the 'medical model' states that the cause of disability and its treatment rest with the individual. In this way the medical model fails to provide a strong basis for the treatment and policy consideration of chronic conditions, which include most disabilities (Smart 2005a, b).

The 'medical model of disability' has had implications for the built environment with specific institutions, rehabilitation centres, hospitals and special schools having been founded to provide specific services for individuals with disabilities. Not only has this model resulted in the exclusion of disabled people from gaining access to education, employment and services, but it has also created social injustice since the built environment and public and private services have presented both architectural and social barriers that have limited integration of those with disabilities into the mainstream society.

According to the 'medical model of disability', individuals with disabilities have been seen as the problem, and so these individuals need to change, adapt and fit in with the physical environment. The social, cultural and physical injustice and exclusion from integration with the mainstream society, in addition to the racial and civil rights and the feminist movements that started in the USA, gave rise to the call by many individuals with disabilities for their equal rights, and the definition of a new model of disability that opposes the 'medical model'. The 'social model of disability' believes that individuals with disabilities are oppressed as a result of the social and architectural barriers that discriminate against, and exclude them from, involvement and participation. As previously mentioned, the social model of disability was developed in the 1970s by activists in the Union of the Physically Impaired Against Segregation (UPIAS). Finkelstein (1980, 1981), Barnes (1991) and Oliver (1990b, 1996) were behind the adoption of the social model of disability, which claims that society discriminates against and restricts individuals with disabilities from accessing public buildings, public transportation, educational institutions and workplace. Oliver (1990b) and Barnes et al. (1999) believe that society prevents individuals with disabilities from integrating with their society and environment. According to the social model of disability, society is the main disabling factor that hinders disabled people from gaining access to services, and those responsible for the facilities of the built environment do not anticipate individuals' physical impairments, and thus do not make decisions to eliminate architectural and social barriers.

In our view, it is society which disables physically impaired people. Disability is something imposed on top of our impairments by the way we are unnecessarily isolated and excluded from full participation in society. Disabled people are therefore an oppressed group in society. To understand this, it is necessary to grasp the distinction between the physical impairment and the social situation, called 'disability', of people with such impairment. Thus we define impairment as lacking all or part of a limb, or having a defective limb, organism or mechanism of the body and disability as the disadvantage or restriction of activity caused by a contemporary social organisation which takes little or no account of people who have physical impairments and thus excludes them from participation in the mainstream of social activities (Oliver 1996, p. 22).

'The social model of disability' calls for the environment to be designed in a way that anticipates the needs of people with disabilities and for their entitlements to gain equal access to buildings, products, services and information, and thus it has played a major and important role in stimulating the adoption of new policies to eliminate social and architectural barriers and promote equality and diversity among the whole population. Goldsmith (1998) distinguishes between the term 'disabled' refereed to the 'medical and the social model of disability'. Whilst the medical model views a disabled person as someone who is incapable of interacting with society as a result of his/her disability and impairment, the 'social model of disability' sees that architects and designers are responsible for creating buildings and spaces that restrict disabled people from accessing and using them. They would not be considered disabled nor subject to discrimination had the relevant buildings been designed to be convenient for everyone (Goldsmith 1998).

Many disability specialists, such as Oliver (1990a, b, 1996) and Barnes et al. (1999), believe that 'the social model of disability' has had a substantial impact on promoting social inclusion, as reflected in the antidiscrimination legislation which has mandated that the built environment should be accessible to disabled people. The Architectural Barriers Act 1968 (ABA) in the USA and the Chronically Sick and Disabled Persons Act 1970 in Britain were the first laws that mandated that buildings and facilities designed, constructed or altered with government funds or leased by a governmental agency should be accessible to disabled people, but they did not address access to governmental programmes and activities that occurred within the buildings or facilities which are not centrally funded (Nussbaumer 2012; Goldsmith 1997). The impact of 'the social model of disability' on the built environment was noted in the production of design guidance and standards associated with the implementation of antidiscrimination legislation and elimination of architectural barriers for people with disabilities (Goldsmith 1997, pp. 101, 121). In 1991, the US Access Board was established to produce accessibility guidelines that would address access to public buildings. In that year, the Board published the first set of Americans with Disabilities Act Accessibility Guidelines (ADAAG) which focused on providing enhanced access for disabled people (Nussbaumer 2012, p. 8). In 1976, British building regulations were introduced for the first time. Part T was the first access guidance designed to eliminate the physical barriers for disabled people in public buildings. However, due to its complexity, it was redrafted in 1987 when it was introduced as Approved Document M, the aim of which was to address accessibility requirements for newly constructed facilities so they would become accessible to, and usable by, disabled people (Goldsmith 1997, p. 77). These guidelines and regulations had limited impact since they focused solely on eliminating architectural barriers for those people with mobility impairments, and thus they were revised to cover a range of facilities in order to address a wide spectrum of disabilities in the public and private sectors (Goldsmith 1997, p. 98).

Although 'the social model of disability' had a substantial impact on promoting social inclusion, many specialists criticised it and called for it to be revised. Finkelstein (2002) claims that since Oliver's social model was specific and radical, it gained an appreciation and general acceptance over the decades. However, nowadays, 'the social model' moved away from the Oliver's radical version and has become associated with the 'rights and legalistic model' of disability, focusing on human rights and equal opportunities.

According to Finkelstein (2002), 'the disabling barrier model' or 'the social model of disability' does not stereotype disabled individuals as aliens or vulnerable individuals who are incapable of functioning, but rather it considers them as citizens who are prohibited from exercising the personal and civil rights that other people are well acquainted with.

Finkelstein (2002) divides the 'disabling barrier model' into two sub-models, the 'restricted citizen model' and the 'dysfunctional model'. 'The restricted citizen model' considers individuals with disabilities whose status is equal to that of all other citizens who are controlled by the environmental and social barriers that have been developed by the able bodied according to their own standards. On the other hand, the 'dysfunctional sub-model', or what is known as 'the medical model', considers impairments and deficits to be purely personal matters which do not result in any social dysfunction.

Associating the 'disabling barrier model of disability' with the care in the community model of intervention triggered the establishment of community care services and independent living centres, whilst the 'dysfunctional sub-model' focused on curing the impairment of disabled people through either surgery or genetics. Finkelstein describes today's 'model of disability' as that of an equal opportunities culture, which encourages individuals with disabilities to assert their individuality within a multicultural society and prompts integration within the able-bodied world.

This model led to the establishment of the independent-living movement in the UK or what Finkelstein (2002) calls the 'independent-living model', which aimed to improve the social services for individuals with disabilities and promote the social model of disability that called for their right to live independently without relying on the welfare and rehabilitation services that were associated with the medical model of disability.

Finkelstein (2002) critiques 'the social model of disability' as it interprets disability in terms of the social context without providing an explanation of the causes of disability. Moreover, 'the social model', as defined by the UPIAS, interprets disability as tragic personal impairment, and believes that society disables people by concentrating on issues concerning emancipation and struggle for social change without acknowledging the changes in cultural lifestyles, individual experiences and their effect on shaping the environment in which the population lives.

Although Finkelstein (2002) gives credit to Mike Oliver for his promotion of the British disability movement's identification of a social model of disability, he believes that Oliver's social model only makes sense when understood in a particular context and at a specific time. In order to construct a new model, one should avoid the artificial separation between the living situation of individuals with disabilities and what he calls the models of disability and the support needed by individuals with disabilities or 'the models of intervention'.

Finkelstein's new model known as the 'diversity in lifestyles model', p. 15, aims to focus on lifestyles rather than needs or services, and it aims to cover the models of disability and the models of interventions that are used by professionals. His proposed model aims to use the individuals' experiences to identify and remove the physical barriers created by and designed for able-bodied lifestyles, attitudes, culture and public transportation. He calls this the 'consultant model'.

His new model urges the development of an inclusive and universal model that takes account of the human social lifestyle and healthcare issues. To achieve such an inclusion, Finkelstein (2002) calls for the repossession of the social model of disability in order to restructure society and make it more diverse by adopting a 'national amelioration service', p. 17, which integrates all services to cover the needs and desires of the whole population and provides services that could improve the quality of living and promote a healthy environment not only for individuals with disabilities, but also for all the population.

On the other hand, Swain and French (2000) call for a new model of disability, 'the affirmative model', p. 580, which is quite similar to Finkelstein's new model. 'The affirmative model' rejects the cultural attitudes and assumptions about seeing individuals with disabilities as victims of personal tragedy, and discards the social model that sees the problem as one that exists within society, but rather associates it with the individuals or their impairments.

'The affirmative model' calls for the inclusion of individuals with disabilities in the mainstream of society and the provision of better physical environments and workplaces that would promote equality and diversity, and a positive image of disabled people, and would accept their difference.

Drawing on Finkelstein's and Swain and French's new models of disability, this research has adopted 'the affirmative model' and has built on it to promote social inclusion, and equality and diversity in educational institutions by acknowledging the new century's demographic changes, lifestyles and market demand to respond to a wide range of users and their different social cultures and ethics, their different age groups and abilities, in order to promote an inclusive university environment from which all users can benefit.

2.2.2   The British Legislation Definition of Disability

Since adopting the social model of disability, British disability organisations and institutions, influenced by the American disability movement, have campaigned for antidiscrimination legislation and civil rights of disabled people. They were behind the first legislation designed to protect disabled people from discrimination in 1970. However, this first legislation, known as the Chronically Sick and Disabled Persons Act, did not define disability. The definition of disability was first introduced when the Disability Discrimination Act (DDA) came into force in 1995, and was amended in 2005.

The DDA 1995 and its amendment in 2005 define an individual with disability as 'a person with a physical or mental impairment which has a substantial and long-term adverse effect on his/her ability to carry out normal day-to-day activities'.

Such a definition has been criticised by many scholars who believe that it reinforces the medical model of disability. Gooding (1996) states that the definition associates 'impairments' with the 'ability to carry out normal day-to-day activities' without correlating it with the social and physical barriers that could prevent a person with disability from interacting with his/her society. Moreover, Gooding (1996) argues that requiring individuals to have experienced disability over a long period of time in addition to the need to provide a proof of the disability has disadvantaged many individuals with disabilities whose impairments did not fall within these criteria.

Intended to promote inclusion and diversity and expand the definition of disability, the UK DDA 2005 was replaced by the Equality Act 2010, which came into force in October 2010. The new Act harmonised the fragmented discrimination legislation; substituted all preceding equality legislation, such as 'the Race Relations Act', 'the Disability Discrimination Act' and 'the Sex Discrimination Act'; and presented a distinctive, united resource of discrimination law. The Equality Act 2010, adopting the inclusive notion, included all types of discrimination that are unlawful in one law by broadening the protection from discrimination to cover nine characteristics: 'age, disability, gender reassignment, marriage and civil partnership, pregnancy and maternity, race, religion and belief, sex, and sexual orientation'. The Equality Act 2010 adopted the same DDA 2005 definition of disability.

> The term 'substantial' means neither minor nor trivial. The term 'long- term' means that the effect of the impairment has lasted or is likely to last for at least 12 months. The term 'normal day-to-day activities' includes everyday things like eating, washing, walking, and going shopping. A normal day-to-day activity must involve one of the capacities listed in the Act, which include mobility, manual dexterity, speech, hearing, seeing and memory (Directgov 2010, p. 1).

Although the current Equality Act 2010 has adopted the same DDA 2005 definition of disability and covers 'chronic diseases such as diabetes, heart disease, and HIV, its connotations with impairment and normal day-to-day activities', as described by Woodhams and Corby (2003, p. 175), it seems to interpret the definition in terms of the 'medical model of disability'.

2.3 Classification and Prevalence of Disability and Implications for the Design of the Built Environment

'Impairment', according to the British Disability Discrimination Act, covers a wide range of impairments, such as sensory impairments affecting sight or hearing, physical impairments, and cognitive and hidden impairments.

2.3.1 Visual Impairment

Visual impairment is defined as a functional limitation of the eye, which reduces its 'visual acuity'. Visual acuity means the clarity or sharpness of vision and the ability of the eye to see and distinguish fine details.

Great Britain uses metric measurements and a typical distance from the patient to the acuity chart is about 6 m. A normally sighted person has a visual acuity of 6/6. He/she can see a particular size of letters at a distance of 6 m. The first number designates the distance from the eye chart, whilst the second number refers to the distance at which the normal-eye-sighted person can distinguish a letter on the same chart clearly.

A person with 'a visual acuity' of 6/18 is only capable of recognising letters at 6 m whilst a normal-eye-sighted person can see the same letters at 18 m. Hence, the level of a person's vision is associated with the increase in the second number of visual acuity. Accordingly, an individual with visual impairment is a person who has a visual acuity in the better eye of less than 6/18 for low vision and in the case of a blind person it is 3/60.

According to the World Health Organization (WHO 2017) there are about 253 million people who are visually impaired; 36 million of them are blind. In the UK and according to the Royal National Institute for the Blind (RNIB 2013), there are two million people or 3% of the total British population who are eligible for registration as blind or partially sighted.

The WHO (2017) states that globally the main causes of visual impairment and blindness are uncorrected refractive errors, which include nearsightedness, farsightedness and astigmatism, and cataract, which is considered to be the prime cause (64%) of blindness (WHO 2017). 'Cataract' occurs when the lens of the eye becomes blurry and prevents lighting from passing through. 'Uncorrected refractive errors' are considered to be the second main cause of visual impairment. 'Glaucoma' is the third main cause and is characterised by the way it damages the optic nerve. The fourth main cause of visual impairment is 'age-related macular degeneration', which leads to loss of the central field of vision. The other main types of visual impairment are 'tunnel vision and diabetic retinopathy'. A person with 'tunnel vision' sees a very small central part of a scene, whilst a person with 'diabetic retinopathy' sees a patchy scene in which objects merge together due to the lack of sharpness across the visual field.

Visual contrast and contrasting colours are a necessity for people with visual impairment. For example a person with only peripheral vision and no colour vision has difficulty identifying fine detail, and the only clues that would assist a person in navigating a space safely and independently would be a change in colour, or visual contrast between wall and floor surfaces that surround a space or environment (Bright and Cook 2010). In the absence of, or in case of only poor, colour contrast in the peripheral field, it is difficult to detect the presence of features unless there is good visual contrast between a wall and the floor, or a door frame and its panel (Bright and Cook 2010).

On the other hand, a person with only central vision, in which the visual field is small, relies on scanning the scene in order to identify the space and the entrance door. Such a person may have relatively good visual acuity, and thus he/she may be able to focus more easily on features such as a door frame and the door handle, and hence providing a colour contrast would help a person to detect the door and its handle (Bright and Cook 2010).

Bright and Cook highlight the importance of providing adequate lighting and visual contrast when designing an inclusive environment. Good lighting provides sufficient illumination to allow users to identify the main features and facilities within a space. Whilst natural light is provided within a space from windows, roof lights and glazed entrances, it is recommended that it be controlled so it does not create glare and confusing shadows that might limit visibility for people with visual impairment.

Moreover, a directional component of lighting can accentuate facial features, which helps people with sensory impairments to lip-read and use sign language. Bright and Cook (2010) stress the importance of supplementing good general lighting of potential hazards, such as steps and ramps, with stronger illumination that permits people with sensory impairments to be aware of such hazards. Whilst contrast between colours is differentiated for most people with good vision because they can discern the nature of colours/hues or the intensity of the colours/chroma, most blind people or those with visual impairments rely on the contrast between colours and the way the light reflects on a surface to obtain information about the space they are in and what it contains. Hence the level of contrast and the value of the reflective light are critical factors when it comes to enhancing or undermining a person's ability to see colour and to identify differences between colours.

According to Bright and Cook (2010), LRV or light reflective value is described in terms of a scale that runs from 0 (zero), a perfectly absorbing surface (totally black), to 100, a perfectly reflecting surface (pure white). The importance of using LRVs as a method of defining appropriate contrast has been included in Approved Document M, which recommends that the difference in LRV between two surfaces be greater than 30 points, based on an illuminance of not less than 100lux.

2.3.2 Hearing Impairment

The Equality Act 2010 protects people with hearing impairment from discrimination. According to the World Health Organization (WHO), 'a person with "hearing impairment" is one who has no hearing at all, or has hearing loss within a particular range of frequencies, or has tinnitus or noise in the ears'. Either people with hearing impairment can be born with a hearing impairment or they can have a health condition or disease which makes it occur later. The hearing ability loss may have implication on the person's ability to talk and communicate when it occurs early. In the UK, under the Equality Act 2010, people with hearing impairments and those who lost their hearing ability completely and are deaf, and also the ones who communicate by using the sign language or those who use hearing aids, are all covered and protected against discrimination. To identify the level of a hearing impairment, a hearing sensitivity test is used to measure the ability of a person to be able to hear sounds and is usually measured in decibels which are units of relative loudness of sounds. A measurement of zero decibels (0 dB) designates the point at which a person with normal hearing ability can sense the faintest sound. A person is considered to be deaf if he/she has difficulty of hearing sounds at the level of about 90 dB or greater, whilst people with hearing impairment at lower decibel levels are considered to be hard of hearing (Hallahan et al. 2015). According to Marschark (1997, p. 41) the Deaf Community in the USA considers people who have some degree of hearing loss due to age to be 'deaf' a with lower case 'd' and does not consider them to be members of the Deaf Community because they can still improve their hearing by using enhancement aids. 'Deaf' (with an upper case initial) is a term that is used to describe people who lost their hearing at a young age. Those people are considered members of the Deaf Community and have their own form of communication separate from that of the hearing world. The Deaf Community's primary means of non-written communication is sign language, which reflects deaf culture and society and is considered to be quite different from hearing people's modes of communication. Marschark (1997, p. 42) notes that 'the Deaf Community has its own social structures, art, clubs and organizations, values and cultural history, which have existed for hundreds of years'.

It is estimated that almost nine million people in the UK have hearing impairments. For deaf people, signing is seen simply as another method of communication, whilst many deaf people with hearing impairments prefer to use induction loops and other technical aids that are provided in public spaces. In addition, there is also an increasing understanding of the importance of sign language and lip reading (Bright and Cook 2010). Whilst the visual contrast and lighting play an important role in making a space usable for people with visual impairment, findings from research undertaken at the University of Reading, known as Project Crystal, stressed that visual contrast and lighting also have an important part in influencing how deaf or hard-of-hearing people gather information, communicate and interact with their surroundings, and the decisions they take when doing so (Bright and Cook 2010).

A key finding from Project Crystal was that well-designed signage that is appropriately positioned is critical for people with the sensory impairment of reduced hearing, just as it is for those with poor vision or for any other disabled person (Bright and Cook 2010). Moreover, whilst the colours chosen for an environment have little influence on how effectively deaf or hard-of-hearing people can communicate within a space, colour contrast between the people involved in the communication and their background is critical. In addition, the nature of the background and whether it provides a plain or busy visual scene also play a vital role in the communication process (Bright et al. 2001).

2.3.3 Physical Disability

According to the definition of disability in the Equality Act 2010, 'a person with physical disability is someone who has a physical impairment which has a substantial and long-term adverse effect on his/her ability to carry out normal day-to-day activities'.

The factors that cause physical disability have been quite varied throughout the nineteenth and twentieth and the beginning of the twenty-first centuries. Many diseases, such as polio and tuberculosis, were the main causes of physical disability in the nineteenth and twentieth centuries. World Wars I and II caused severe injuries and disabilities in the twentieth century throughout the whole world. By the end of World War II, professionals, engineers and scientists were focusing on developing new medical treatments and facilities to improve the quality of life. Creating vaccines against polio and tuberculosis had certainly prevented some of the causes of physical disability; however, the number of brain and spinal injuries arising from the use of motor cars and motor cycles caused the number of people with physical disabilities to increase. There are many diseases that can cause physical disability. Some of these diseases can be inherited or caused by a genetic disorder. Diseases such as muscular dystrophy, spina bifida, meningitis and other illnesses that can affect the brain and the human nerves system and muscles can lead to physical disability.

The two main types of physical disability that are acquired as a result of a disease are 'multiple sclerosis' and 'polio' which are caused by inflammatory diseases that affect the central nervous system. On the other hand, muscular dystrophy is a genetic disease that affects the body muscles and limits a person's ability to move and walk. Cerebral palsy, on the other hand, is the result of damage to the brain tissue of the foetus.

Whilst visual contrast and lighting play an important role in making a space usable for both people with visual and hearing impairments, level access, passenger lifts, comfortably graded steps and stairs equipped with handrails on both sides, and spacious public toilets that can accommodate a wheelchair and are conveniently reachable by wheelchair users are important provisions for people with mobility impairments (Goldsmith 2001, p. 2).

2.3.4 Cognitive Impairments/Learning Difficulties

The World health Organization (WHO) defines learning disabilities 'as state of arrested or incomplete development of mind'. According to Hallahan et al. (2014, p. 168), in the USA 'the two most influential definitions of learning disabilities are the federal one and that of the National Joint Committee on Learning Disabilities (NJCLD) definitions'. The federal definition uses the term 'specific learning disability', describing it as

> A disorder that can include one or more of the basic psychological processes involved in understanding or in using language, spoken or written; such a disorder may manifest itself in an imperfect ability to listen, think, speak, read, write, spell or do mathematical calcula-

tions. It can include perceptual disabilities, brain injury, minimal brain dysfunction, and dyslexia and development aphasia (Hallahan et al. 2014, p. 168).

On the other hand, the National Joint Committee on Learning Disabilities (NJCLD), dissatisfied with the federal definition, proposed a different definition. The NJCLD defines learning disabilities as

> A term that refers to a heterogeneous group of disorders, manifested by significant difficulties in the acquisition and use of listening, speaking, reading, writing, reasoning or mathematical abilities. These disorders may affect an individual's intellectual abilities due to central nervous system dysfunction, and may occur across the lifespan (Hallahan et al. 2014, p. 169).

As Hallahan et al. (2014, p. 169) affirm, 'An individual with a learning disability is a person who has a significant impairment of intellectual functioning and of adaptive or social functioning'. This affects the ability of a person to process information, learn, understand, write and remember new things, and use the knowledge to solve problems. Since learning is essential to overcome problems and barriers, people with learning difficulties encounter problems that may restrict them from interacting and socialising with others, managing their daily activities independently and safely.

There are many terms that have been used by scholars to define a person with cognitive impairment or learning difficulties. The most commonly used terms are learning difficulties, dyslexia, specific learning disabilities, specific learning difficulties, word blindness and specific reading retardation (Al-Hroub 2007, 2010, 2013; Hornsby 1994). Whilst physical disabilities can be clearly distinguished and acknowledged due to the physical impairment that may restrict a person from interacting with his/her physical environment, 'learning difficulties' or 'dyslexia' has been labelled as 'hidden disabilities' by many scholars and educators since many of them were unable to differentiate between the person's intellectual ability and academic achievement. According to the British Dyslexia Association (Crisfield 1996), and the Warnock Report (DES 1978), specific learning disabilities/difficulties are defined as a discrepancy between a person's intellectual ability and his/her school academic achievement (Al-Hroub 2006).

As far as the built environment is concerned, confusing routes that are hard to navigate in buildings create a major barrier for people with learning difficulties. According to the Approved Document M, clearly signposted spaces with adequate visual contrast and lighting can provide an easy-to-navigate space from which all users can benefit, including those with learning difficulties.

2.3.5   Autism Spectrum

Autism and Asperger's syndrome are considered to occupy different parts of the same autistic spectrum (Harker and King 2002, p. 10). In the UK, the National Autistic Society (2012) defines autism as 'a lifelong developmental disability that affects how a person communicates with, and relates to, other people'. Autism, as defined by the Individuals with Disabilities Education Act (IDEA) in the USA, is 'a developmental disability affecting verbal and non-verbal communication and social interaction, generally evident before the age of 3, which affects a child's performance' (Hallahan et al. 2014, p. 280). Some individuals with autism manage their daily activities independently, whilst others may need special support and assistance since they may have learning difficulties. Asperger's syndrome is a mild type of autism. 'A person with the Asperger syndrome has an average or above-average intelligence. He/she often shows inappropriate social behaviour, circumscribed interests, and show awkwardness of speech, lack of common sense and specific learning disabilities' (Harker and King 2002, p. 11). Individuals with autism may also be sensitive to sound, touch, taste, smell, light or colour, and hence such specific needs should be considered in the built environment they use.

This study aims, having reviewed the different categories and kinds of sensory impairment and physical and learning disability, to investigate whether universities have managed to accommodate all these needs to improve the accessibility of their built environments.

2.4   The Disability Rights Movement

Throughout history individuals with disabilities and impairments have suffered from discrimination and marginalisation. Society has considered individuals with disabilities as abnormal persons and has thus treated them with pity, sympathy or fear. According to Hutchison (2002), in Scotland individuals with disabilities were seen as different and thus regarded as 'other' during the nineteenth century, a time when the notion of institutions was growing and expanding. Thus individuals with mental and sensory impairments were placed in special institutions and excluded from the family and the local community. Campbell and Oliver (1996) affirm that in the nineteenth century Britain and most Western societies deemed individuals with disabilities to be inadequate individuals who were unable to take part or function in the mainstream society alongside able-bodied individuals.

The disability rights movement throughout history has taken different paths and approaches to claiming the rights of people with disabilities in different countries. Accordingly, this section reviews the disability rights movement in Britain.

2.4.1   The Disability Rights Movement in Britain

In the nineteenth century individuals with disabilities in Britain were generally thought of as incapable of functioning independently and had to depend on others to carry out their day-to-day activities. This led to the spread of charitable foundations that aimed to support and help them both morally and financially. By the end of the Second World War, the British Government had introduced the welfare state to improve the infrastructure and facilities for the whole population. Campbell and Oliver (1996) state that the first legislation relevant to individuals with disabilities and their needs was the Disabled Persons Act 1944. However, successive governments failed to enforce it and the legislation has been amended and replaced many times during subsequent decades. Table 2.1 summarises the history and evolution of the British disability rights movement and legislation.

2.4.2   The UN Convention

In 1948 the United Nation produced the Universal Declaration of Human Rights and in 1975 the United Nations Declaration on the Rights of Disabled Persons was adopted by the UN General Assembly. The declaration was based on the 'rehabilitation model of disability', which perceived individuals with disabilities as a group of people who needed special provisions, facilities and support, and acted upon that consideration. The disability rights movement and 'the social model of disability' played a major role in changing the mind set about people with disabilities. This was revealed in the amendment of the Convention on the Rights of Persons with Disabilities and its Optional Protocol in 2006. The UN Convention aimed to integrate individuals with disabilities in their societies by outlawing discrimination and setting obligations and requirements to remove barriers to enable people with disabilities to gain access to facilities and services in their societies. In May 2008 both the UN Convention and its Protocol were introduced. It emphasised the necessity of mainstreaming disability issues as an integral part of relevant strategies used to promote sustainable development. The UN Convention also stressed the importance of improving the accessibility to the physical, social, economic and cultural environment; to health and education; and to information and communication to enable persons with disabilities to fully enjoy all human rights and fundamental freedoms. It also emphasised the importance of international cooperation for improving the living conditions of persons with disabilities in every country, particularly developing countries.

According to the UN Convention (2006), 135 states have signed the Convention, 40 states have also ratified it, 75 have signed it and 24 have ratified the Optional Protocol. All the EU member states, including the UK, have signed the Convention.

2.4.3   Inclusive Education

Prior to the anti-discrimination disability and civil rights legislation, disability as defined according to the 'medical model' prohibited many individuals with disabilities from gaining access to schools, facilities and services in their communities. Their impairment was considered as the main disablement which hindered them

Table 2.1 Historical review of the British disability rights movement and legislation

Legislation	
Disabled Persons Act (1944)	• Affirmed that facilities and services had to be reasonably accessible to disabled employees.
Education Act (1944)	• Recommended the inclusion of students with disabilities in ordinary schools.
	• The Act could not be implemented since many special schools and programmes were being offered to students with disabilities, resulting in their segregation.
National Health Services Act (1948)	• Marked the establishment of many care hospitals, where individuals with disabilities were segregated in special wards.
National Assistance Act (1948)	• Gave local authorities the right to provide either community-based or residential services for individuals with disabilities who had either to live in the community and manage with no services or to stay in special wards or residential institutions.
Disabled activists actions (1960s)	• Influenced by the civil rights movements, the women's movement and the independent living movement in the USA, many disabled activists in the UK ran their own local disability organisations such as the Association of Disabled Professionals, established in 1971.
British Standard Code of Practice CP96 on access for the Disabled to Buildings (1967)	• In 1962 a British Standards Institution Committee was established to produce a national access standard based on the American National Standard A117.1, published in 1961.
	• Selwyn Goldsmith, an architect who became disabled by polio, was in charge of drafting the standard, which was issued in 1961.
	• Unisex accessible toilets were first introduced requiring an internal dimension of 1370 × 1750 mm.
Chronically Sick and Disabled Persons Act (1970)	• Set out special provisions for individuals with disabilities and treated them differently.
	• Contained 29 sections.
	• Section 3 placed duties on local authorities to make new housing provisions accessible to individuals with disabilities. It also included provision of specially designed housing suitable for wheelchair users to assist housing authorities to comply with the Act.
	• Section 4 focused on providing access to public buildings and required that the providers of public buildings make their premises accessible.
	• Did not provide any tools to enforce the Act in practice.
	• The equal rights of disabled people to gain access to education were not stated in the 1970 Act. Disabled people were still excluded from gaining access to public education. Special schools were only provided for disabled people, which hindered their integration into society.

(continued)

Table 2.1 (continued)

Legislation	
	• There were no penalties for non-compliance with Section 4 requirements incorporated within the Act.
	• There was no order to produce regulations for the enforcement of Section 4.
	• The Act did not specify which government department was to be responsible for putting Section 4 requirements into operation.
	• The Act did not define the term 'any provisions' and whether these provisions should be applied to new constructions or alterations to existing buildings.
Minister for the Disabled (1974)	• Alf Morris became the first Minister for the Disabled.
British Building Regulations, 1976	• The construction of buildings was associated with the health and safety requirements of the Public Health Acts of 1936 and 1961.
	• In 1976, the British Building Regulations were published as a substantial package of technical prescriptions specifying precisely what had to be done for compliance purposes (Goldsmith 1998, p. 76).
	• Part T was the first stage of a legislative process for enforcing access requirements and was an add-on to the 1976 Building Regulations.
The social model of disability adopted by the Union of the Physically Impaired Against Segregation (UPIAS), 1970s	• Oliver, Finkelstein, and Barnes were disability activists who were behind the adoption of the social model of disability in the early 1970s. The social model of disability claims that society discriminates against and restricts individuals with disabilities when it comes to accessing facilities and services as people in charge of shaping the built environment do not anticipate disabled people's physical needs or remove architectural and physical barriers. The impact of the social model of disability was reflected in the disability anti-discrimination legislation and design guidelines which called for the elimination of architectural barriers to disabled people.
British Council of Disabled People (BCODP) (1981)	• Established by individuals with disabilities to promote their full equality and participation within society.
Disabled Persons Act (1982)	• Issued after reinserting the Building Regulations provisions of 1976.
	• Required that a building provider comply with the appropriate provisions and the design codes of practice when designing a building in order to make it accessible for individuals with disabilities.
	• Educational buildings were not included in the Act and thus they were inaccessible to disabled people.

(continued)

Table 2.1 (continued)

Legislation	
Building Regulation 1985	• The Building Regulations (1976), with their inflexible requirements based on traditional construction methods, discouraged innovative techniques, and many professionals and architects called for a recasting of the regulations.
	• In 1982, the decision was made to simplify the building regulation standards so that they should contain a series of functional requirements with the additional stipulation that the building shall not catch fire, give off harmful fumes or smell.
	• The recasting of building regulations meant that these requirements would be issued in parts and each part would have its Approved Documents.
	• The purpose of Approved Documents was to provide guidance on how to satisfy the regulation requirements.
	• The assembling of Approved Documents had already been done by 1983 and it was too late to add the access regulations, which needed to be prepared and drafted.
	• The recast regulations in 11 parts from A to L became known as Building Regulation 1985.
	• BS5810 'Code of practice for access for the disabled to buildings' was introduced as the fourth amendment to the 1976 Building Regulations, which at that time were in parts from A to S. It dealt only with design standards and did not cover application conditions.
	• It was proposed that Part T, concerning access for disabled people, should be drafted so it could be added to the 1985 Building Regulations.
	• In acknowledgment that Part T was of particular concern to disabled people whose interests were represented by disability organisations, it was placed under the welfare and convenience remit of the 1974 Health and Safety at Work Act.
	• In February 1983 consultations took place to draft the new Part T, based on provisional conditions that new public and employment buildings would be required to be accessible if they were above a certain size and that alterations or extensions to existing buildings would not be covered. In October 1983, the size criterion for new buildings was dropped and only certain alterations and extensions were to be covered.
	• Part T was based on BS5810 that specified design standards for access for the disabled, but there was no related code of practice which prescribed suitable provision of exits for the disabled if they were in a multistorey building and without an exit code. Part T required that only single-storey buildings had to be accessible to disabled people.

(continued)

Table 2.1 (continued)

Legislation	
	• 'Part T access regulations were then added to BS5810, and were tackled in Schedule 2 of the amended 1985 regulations. Item T2 Provision of facilities for disabled people decreed that the regulation applied to (a) office and shop buildings, (b) single-storey buildings, (c) single-storey school buildings, and (d) other single-storey buildings if they were buildings that were used by the public, whether or not requiring payment for access. Enforcement relied on BS5810' (Goldsmith 1997, p. 88).
	• It was hoped that Part T would be a breakthrough and would achieve normalisation and a shift in attitudes towards the macro approach to anticipating the needs of all users, including disabled people; however, this hope could not be realised.
	• It was hoped that normalisation would be achieved with the production of a new Approved Document, named Part M, which was put out for consultation in September 1986. However, the majority of committee members in charge of drafting Part M favoured following the micro approach, namely making specific provisions to cater for disabled people's specific needs.
	• The 1985 regulations were amended in 1987 after the Approved Document was drafted with the access regulations incorporated as Part M.
	• 'The 1987 Part M regulation, like the 1985 Part T, was minimalistic in form and applied to only new multistorey buildings used as offices and shops. The number of lifts provided in new multistorey buildings was revised in Part M and the number of lifts was determined by floor area' (Goldsmith 1997, p. 98).
	• Whilst Part T 1985 regulated single-storey buildings, Part M 1987 required that the main entrance of all new multistorey public buildings should be accessible. To comply with Part M, any new building was subject to M2, M3 and M4; M1 contained definitions.
	• M2 contained provisions concerning the means of access to all new buildings; M3 contained requirements relating to sanitary provision; and M4 contained requirements relating to seating provision for audiences or spectators. All of these provisions focused on providing access for people with physical impairments and wheelchair users without taking into account the needs of people with different disabilities.
	• Part M 1987 was ambiguous and only focused on catering for people with physical impairments. A decision was made to revise and improve it following the issuing of a code of practice on exit for people with disabilities.
	• In 1988 the British Standards Institution issued BS5588 'Fire precautions in the design and construction of building'. Code of practice 'for means of escape for disabled people'.
	• In 1990, the Department of the Environment called for the revision of Part M, which became the 1992 Part M Regulation.

(continued)

Table 2.1 (continued)

Legislation	
	• 'Part M Regulation came into force in 1992 and aimed to provide access to people with physical, visual and hearing impairments. It covered existing buildings, extensions to existing buildings and new constructions and was applicable to all storeys in new and non-domestic buildings' (Goldsmith 1997, p. 114).
Disability Discrimination Act (1995)	• This Act outlawed discrimination against individuals with disabilities in relation to employment, the provision of goods and services, renting or buying land or property, education and transport.
	• Prior to 1995, education was excluded from the DDA.
	• Although the DDA 1995 does not directly require buildings to be accessible to all disabled people and does not include standards for accessible building design, it is the services on offer within educational or any other buildings that are the concern of the Act (Sawyer and Bright 2007, p. 5)
	• Prior to the DDA 1995, disabled people seeking to pursue their education had no other option than to enrol in special schools that segregated them from society. With the introduction of the DDA 1995, education was covered under the Act and education providers had duties to enhance their services to accommodate disabled people.
	• Before introducing the DDA 1995, local authorities through planning controls and the Part M regulations had the option to follow the British legislation concerning the accessibility of buildings. The DDA 1995 changed this principle and gave the disabled person the power to sue organisations or service providers of buildings if they failed to eliminate barriers or make reasonable adjustments for people with disabilities to gain access to such services or buildings.
	• Part I contained the definition of disability that covered those with physical and mental impairments, such as people with mobility, visual, hearing and cognitive impairments.
	• Part II focused on employers, placing duties on employers not to discriminate against disabled people in their employment or when such people are applying for a job. Under Part II employers have to make reasonable adjustments to eliminate the physical barriers on their premises or any arrangements that can cause a substantial disadvantage to a disabled employee or job applicant.
	• Part III focused on access to goods, facilities, services and premises. It placed the duty on service providers not to discriminate or treat disabled people less favourably because of their disability.
	• To eliminate physical barriers, service providers have to comply with building regulation standards, such as Part M or British Standards.
	• Education Act 1996: This raw with the one below define the Education Act 1996. To implement the DDA 1995 in the educational sector, the Education Act was introduced in 1996. It aimed to address the educational needs of children with special needs.

(continued)

Table 2.1 (continued)

Legislation	
	• The DDA 1995 and the Education Act 1996 provided the statutory framework that underpins the equal rights of disabled students to gain access to education.
	• Special Education Needs and Disability Act (2000) SENDA: Was endorsed in 2002.
	• Placed duties on universities, colleges and local education authorities not to discriminate against learners with disabilities by providing reasonable adjustments and accommodations to cater for their needs.
British Standards BS 8300:2001	• BS 8300:2001 'Design of buildings and their approaches to meet the needs of disabled people'. 'Code of practice' was published in 2001 and was amended in June 2005.
	• 'BS 8300:2001 is a research-based document which looks into ergonomic issues, such as reach ranges and space requirements, in order to assess the accessibility for people with disabilities. It provides detailed guidance on the design of domestic and non-domestic buildings, and covers the environmental needs of people with disabilities' (Sawyer and Bright 2007, p. 5).
	• BS 8300:2001 includes provisions for car parking, access routes, entrances and interiors, horizontal and vertical circulation, surface finishes and communication aids.
	• BS 8300:2001 is a comprehensive standard which can be used by architects and designers as an accessible benchmark that includes accessible design provisions and solutions that are accessible to people with disabilities.
The 2004 edition Approved Document M (2004)	• After BS8300 was published in 2001, the Part M Building Regulation standard was revised and a new edition was published in 2004. The new edition named Approved Document M 'Access to and use of buildings' reflects the change in attitude towards including provisions that take into consideration the diverse needs of users (wheelchair users, people with visual and hearing impairment).
	• The Approved Document M provisions and dimensions have been amended and revised so that it complies with the BS 8300:2001 design guidance (Sawyer and Bright 2007, p. 5).
Disability Discrimination Act (2005)	• It has amended the DDA 1995 definition of disability and has brought the requirements used to establish a mental impairment in line with those employed to establish a physical impairment.
	• People with chronic diseases such as those with HIV, cancer, diabetes and multiple sclerosis are all protected against discrimination under the Act.

(continued)

Table 2.1 (continued)

Legislation	
	• The design standards prescribed in Part M 1992 and the changes made in 1999 have been criticised for following the micro or the top-down approach which anticipates the needs of disabled people but not those of other users who might encounter physical barriers, such as children, women and old people.
	• Education is covered by the DDA 2005. It requires all educational providers to make reasonable adjustments to eliminate accessibility barriers and provide special educational programmes for students with special needs.
Equality Act 2010	• It came into force in October 2010.
	• It puts an obligation on employers, service providers, those selling or letting land and property, and education providers to make reasonable adjustments to eliminate architectural barriers.
	• It outlaws discrimination against students at education institutions, and universities due to their sex, race, disability, religion or belief and sexual orientation.
	• It places duties on education providers to remove barriers and make reasonable adjustments for staff and students to enable them to gain access to facilities, services and information and thus provide auxiliary aids for students with special needs or learning difficulties.
BS8300-2010 and Approved Document M-2010	• In 2009 a new edition of BS8300 was published and in 2010 Part M was revised and a new edition was published to incorporate the inclusive design approach and address the needs of a wide range of users with disabilities.

from interacting with the society. Individuals with disabilities were excluded from gaining access to formal education in both developed and developing countries.

According to Jarret (2012), such negative attitudes towards individuals with disabilities has led to the establishment of segregated institutions that first started in the nineteenth century and continued in the twentieth century, where people with disabilities were considered different and had to be placed in separate special schools.

Many laws were endorsed between the two World Wars that promoted the notion of exclusion and segregation among people with disabilities, particularly people with learning difficulties. Such laws considered people with disabilities as second-class citizens and proposed providing special schools for students with learning disabilities that separated them from mainstream education.

Special schools were first introduced in the fifteenth century and belonged to private charitable organisations. The first such schools were dedicated to individuals with sensory impairments and then developed into those that catered for the remaining disability groups. Moreover, the special schools' curriculum was different from that of mainstream schools as it focused on teaching vocational skills. In the 1930s, the IQ test was developed and students obtaining low grades in the test were withdrawn from regular schools and even restricted from enrolling at special schools.

In 1944, the Education Act (1944) was introduced in the UK. It recommended that individuals with disabilities are integrated in regular schools; however, the Act could not be implemented as many private special schools were still hosting students with disabilities.

After the World War II, the National Health Service was introduced in 1946 in the UK, and many hospitals were built to host British veterans and workers who were severely injured by work accidents and needed healthcare services and rehabilitation centres. Consequently, during the same period 'the medical model of disability' was developed and physical and mental disability was perceived as the main disablement that hinders a person from interacting with his/her society. Hence people with physical disabilities were placed at hospitals and rehabilitation centres, whilst people with mental and learning disabilities had to attend special institutions and schools.

The disability right movement which started in the 1960s and 1970s changed the negative attitude towards people with disabilities. People with disabilities rejected to be treated differently and be placed at segregated institutions and hospitals and called for their equal civil rights. They demanded to be treated equally and called for their right to gain access to education and employment.

After the adoption of the Universal Declaration of Human Rights after the World War II, the right to education became mandatory for all citizens. Article 26 of the Declaration proclaims that citizens have equal right to adequate education, despite their gender, race, colour and religion.

The Universal Declaration of Human Rights, along with the removal of institutionalisation, has led to the concept of normalisation in the educational system that aims to integrate individuals with disabilities into mainstream society and culture. Moreover, normalisation in the educational system focuses on maximising the use of the regular school system and minimising the use of separate or special educational facilities.

Although the Declaration has placed duties on national governments to guarantee the right to education for all citizens, many people with disabilities have still been marginalised and educated in special educational systems as their individual differences have not been recognised.

This condition has motivated many individuals with disabilities to revolute against these discriminatory actions and called for their equal rights to be able to contribute to their society. They called for their equal right to be able to access and choose the schools, education institutions, workplaces and public transportation.

The disability right movement which started in the USA had an influence in enhancing accessibility at educational institutions and the University of Illinois Champaign-Urbana was the first US university which had accessibility provisions for people with disabilities.

Advocates of accessibility and architects in the USA to work on eliminating physical barriers, and the University of Illinois Champaign-Urbana was the first university in the world to attempt to enhance the accessibility of its buildings to disabled people in 1956. Moreover, the independent living movement, which started in the USA at the University of California Berkeley in 1972, called for the removal of architectural and transportation barriers that prevent people with disabilities from having equal opportunity to share fully in all aspects of society (Goldsmith 1997, pp. 9, 53–56).

The demand for establishing independent living centres which started at the University of California Berkeley had led to the establishment of hundreds of independent living centres and units across the USA, and many other countries including the UK.

> This movement along with disability rights and access advocacy has called for the need to design buildings accessible to people with disabilities and raised the awareness of the need to bring people with disabilities into mainstream society, ensuring equal opportunity and eliminating barriers to access to, and use of, the built environment (Steinfeld and Maisel 2012, p. 15).

The disability rights movement has played a major role in the recognition of the importance of inclusive education and inclusive design. Many disability groups and activists have perceived disability as a social barrier that prevents them from integrating into their societies, excluding them from receiving appropriate education (Oliver 1990b). Moreover, many of them have perceived integration as only providing additional arrangements to accommodate their physical needs within a school system that nevertheless remains largely unchanged. Hence they have moved from the concept of integration towards that of inclusive education and inclusive schooling.

Inclusive education is a process that involves changing attitudes, policies and practices by restructuring school settings in order to anticipate the physical and learning needs of all students. Whilst integration calls for separate arrangements in the regular school for students with disabilities and those with learning difficulties, by either withdrawing them from regular classes or providing remedial education, inclusive education acknowledges that special learning needs can arise from social, psychological, economic, verbal, cultural and physical factors. Hence to anticipate the individual differences and needs of diverse students who might experience difficulty in learning, inclusive education recognises the necessity to continually review the school system to meet the needs of all learners (Al-Hroub 2009).

The Regular Education Initiative (REI), which started in the USA, called for the adoption of inclusive education by merging special and general education into one single educational system that all students with different abilities could attend. To achieve that inclusion, special education staff and resources were recruited to integrate students with different abilities into the mainstream education (Skrtic 1991).

In the UK, education was excluded from the disability legislation until 1995. The DDA 1995 was the first antidiscrimination legislation which prohibited the exclusion of disabled people from gaining access to education. The Special Educational Needs and Disability Act 2001 (SENDA) was introduced in 2002 and prohibited discrimination against individuals with disabilities in education, training and any services provided wholly or mainly for students at schools, higher education institutions and sixth-form colleges.

Under the Act, students with disabilities have equal rights in terms of having access to educational services, such as field trips, examinations and assessments, short courses, arrangements for work placements, and libraries and learning resources. It is considered unlawful to treat an individual with a disability less favourably than a non-disabled person or prohibit him/her from enrolling at a school or university. Schools and universities are requested to change their policies and practices, course requirements or work placements and the physical features of a building to accommodate the needs of students with disabilities. Moreover, the Act places duties on education institutions to provide interpreters and other support workers, and auxiliary aids and materials in different formats, and to deliver courses in alternative ways to accommodate the needs and learning styles of students with disabilities. In addition, the Act places duties on educational organisations to ensure that their websites are constructed so as to be accessible and easy to navigate for all users, including people with visual and hearing impairments.

In addition to the SENDA, codes of practice were introduced to guide schools and higher education institutions in how to promote inclusive education and educational services in order to tackle the needs of individuals with disabilities.

In 2010, the Equality Act was introduced and placed duties on schools and universities not to discriminate against students because of their sex, race, disability, religion or belief, or sexual orientation (Equality Act, 2010). The new legislation placed an obligation on education providers to make reasonable adjustments to eliminate physical and learning barriers and enhance provisions, criteria and practices in order to ensure that individuals with disabilities will receive the same quality of service as that available to other non-disabled people.

The book focuses on the social model of disability and on human civil rights in order to promote inclusive and equitable education. It supports the concept of inclusion in both physical and educational environments in order to accommodate the needs of all learners and maximise their participation.

To achieve this approach in university settings, the study presented in this book aims to acknowledge the challenges and barriers that restrict universities in the UK when it comes to inclusiveness by drawing attention firstly to the evolution of the educational system in the UK. Chapter 3 then reviews the emergence of the standard codes and how they are reflected in the physical environment.

References

Al-Hroub, A. (2006). *Identifying and programming for mathematically gifted children with learning difficulties*. Unpublished PhD dissertation. University of Cambridge.

Al-Hroub, A. (2007). Parents' and teachers' contributions to identifying the unusual behavioural patterns of mathematically gifted children with learning difficulties (MG/LD) in Jordan. *The Psychology of Education Review, 31*, 8–16.

Al-Hroub, A. (2009). Dynamic assessment applied to preschool children with learning difficulties. *La Nouvelle revue de l'adaptation et de la scolarisation, 46*, 61–76.

Al-Hroub, A. (2010). Perceptual skills and Arabic literacy patterns for mathematically gifted children with learning difficulties in Jordan. *The British Journal of Special Education, 37*, 25–38.

Al-Hroub, A. (2013). Multidimensional model for the identification of gifted children with learning disabilities. *Gifted and Talented International, 28*, 51–69.

Barnes, C. (1991). *Disabled people in Britain and discrimination*. London: Hurst and Co.

Barnes, C., Mercer, G., & Shakespeare, T. (1999). *Exploring disability: A sociological introduction*. Malden: Blackwell Publishers.

Bright, K., & Cook, J. (2010). *The colour, light and contrast manual -designing and managing inclusive built environments*. Oxford: Wiley Blackwell.

Bright, K. T., Cook, G., & Luck, R. (2001). *Project Crystal: Enlightening communication*. Paper presented at COMRA 2001. The construction research conference of the Royal Institute of Chartered Surveyors Foundation, 23–24 June 2001, Strathclyde, Glasgow. Retrieved July 20, 2018, from https://www.wiley.com/legacy/wileychi/brightandcook/supp/scbil2001b.pdf

Campbell, J., & Oliver, M. (1996). *Disability politics: Understanding our past, changing our future*. London: Routledge.

Crisfield, J. (1996). *The dyslexia handbook 1996. Reading*, United Kingdom: British Dyslexia Association.

DES (1978). *Special educational needs. Report of the committee of enquiry into the education of handicapped children and young people* (Warnock Report). London HMSO.

Directgov. (2010). *Definition of 'disability' under the Disability Discrimination Act (DDA)*. Retrieved July 20, 2018, from http://www.lancashire.gov.uk/media/898536/disability-discrimination-act-2005-definition.pdf

Education Act. (1944). *Education Act 1944: Chapter 31*. Retrieved July 30, 2018, from http://www.legislation.gov.uk/ukpga/Geo6/7-8/31/enacted

Finkelstein, V. (1980). *Attitudes and disabled people*. New York: World Rehabilitation Fund.

Finkelstein, V. (1981). To deny or not to deny disability. In A. Brechin et al. (Eds.), *Handicap in a social world*. Hodder and Stoughton: Sevenoaks.

Finkelstein, V. (2002). *The social model of disability repossessed*. Coalition, February, 10–16.

Goldsmith, S. (1997). *Designing for the disabled, The new paradigm*. Oxford: Architectural Press.

Goldsmith, S. (1998). Delays to disabled access. *The Architect's Journal, 208*(9), 63–65.

Goldsmith, S. (2001). *Universal Design. A manual of practical guidance for architects*. Oxford: Architectural Press.

Gooding, C. (1996). *Blackstone's guide to the Disability Discrimination Act 1995*. London: Blackstone Press.

Hallahan, D., Kauffman, J., & Pullen, P. (2014). *Exceptional Learners: An introduction to special education* (12th ed.). Upper Saddle River, NJ: Pearson.

Hallahan, D., Kauffman, J., & Pullen, P. (2015). *Exceptional Learners: An introduction to special education*. Boston: Pearson.

Harker, M., & King, N. (2002). *Designing for special needs: An architect's guide to briefing and designing options for living for people with learning disabilities*. London: RIBA Publishing.

Holmes-Siedle, J. (1996). *Barrier-free design: A manual for building designers and managers*. Oxford: Butterworth-Heinemann.

Hornsby, B. (1994). *The Hornsby correspondence course; Module 1-4* (5th ed.). London: The Hornsby International Centre.

Hutchison, I. (2002). Disability in nineteenth century Scotland: The case of Marion Brown. *University of Sussex Journal of Contemporary History, 5*, 1–18.

Jarret, S. (2012). Disability in time and place, *English Heritage disability history web content*. Retrieved July 30, 2018, from https://content.historicengland.org.uk/content/docs/research/disability-in-time-and-place.pdf

Kaplan, D. (1998). *The definition of disability*. Retrieved August 6, 2018, from http://www.accessiblesociety.org/topics/demographics-identity/dkaplanpaper.html

Marschark, M. (1997). *Looking beyond the obvious: Assessing and understanding deaf learners*. Rochester, NY: National Technical Institute for the Deaf. Rochester Institute of Technology.

Nussbaumer, L. (2012). *Inclusive design: A universal need.*. New York: Fairchild; London: Bloomsbury.

Oliver, M. (1990a). *Politics and language: The need for a new understanding* (Disability, citizenship and empowerment. K665, Workbook 2, Appendix 4). Milton Keynes: Open University.

Oliver, M. (1990b). *The politics of disablement*. Basingstoke: Macmillan.

Oliver, M. (1996). *Understanding disability: From theory to practice*. Basingstoke: Macmillan.

Pfeiffer, D. (1998). The ICIDH and the need for its revision. *Disability and Society, 3*(4), 503–523.

Riggar, T. F., & Maki, D. R. (2004). *Handbook of rehabilitation counselling*. New York: Springer Publishing.

RNIB. (2013) *Number of adults and children certified with sight impairment and severe sight impairment in England and Wales: April 2011–March 2012*. RNIB.

Sawyer, A., & Bright, K. (2007). *The access manual: Auditing and managing inclusive built environments*. Oxford: Blackwell Publishing Inc..

Skrtic, T. (1991). *Behind special education: A critical analysis of professional culture and school organization*. Denver: Love Pub. Co..

Smart, D., & Smart, J. (2006). Models of disability: Implications for the counseling profession. *Journal of Counseling and Development, 84*(1), 29–40.

Smart, J. F. (2005a). Challenges to the biomedical model of disability: Changes to the practice of rehabilitation counseling. *Directions in Rehabilitation Counseling, 16*(4), 33–43.

Smart, J. F. (2005b). The promise of the International Classification of Functioning, Disability, and Health (ICF). *Rehabilitation Education, 19*(2/3), 191–199.

Steinfeld, E., & Maisel, J. (2012). *Universal design: Creating inclusive environments*. Hoboken, NJ: John Wiley & Sons.

Swain, J., & French, S. (2000). *Towards an affirmation model of disability. Disability & Society, 15*(4), 569–582.

The Architectural Barriers Act. (1968). Architectural Barriers Act (ABA) of 1968. Retrieved July 20, 2018, from https://www.access-board.gov/the-board/laws/architectural-barriers-act-aba

Woodhams, C., & Corby, S. (2003). Defining disability in theory and practice: A critique of the British Disability Discrimination Act 1995. *Journal of Social Policy, 32*(2), 159–178.

World Health Organisation (1980). *International classification of impairments, disabilities and handicaps*, WHO, Geneva.

World Health Organisation. (2017). *Blindness and vision impairment prevention: 2017*. Retrieved August 6, 2018, from http://www.who.int/blindness/en

Chapter 3
Emergence of Design Standards and Inclusive Design

3.1 The Emergence of Standard Codes

Acknowledging that legislation in the UK places duties on education institutions and providers to remove architectural barriers and promote inclusion in the mainstream, this chapter aims to investigate whether universities in the UK have complied with the legislation to promote inclusion for all potential users, including individuals with disabilities. The study, adopting the inclusive design principles and approach, aims to examine how the legislation has influenced building design standards and regulations to become more inclusive. Hence this chapter reviews the evolution of the building regulation standards in Britain.

3.1.1 The Emergence of the Standard Codes of Practice in Britain

Influenced by the American disability rights movement, Britain followed the same path in dealing with issues that concern individuals with disabilities. Goldsmith (1998) points out that Britain in the early 1960s was not aware of the needs of such individuals and thus had no programmes for making public buildings accessible to them. Whilst Nugent was the American pioneer in setting standard design codes for individuals with disabilities, Goldsmith was the progenitor of the British Access Standards Institute code of practice that required unisex public toilets.

At the age of 23 Goldsmith was infected by the polio virus, which meant that he became a person with a disability. In 1961 Duncan Guthrie, the Director of the Polio Research Fund, encouraged Goldsmith to write a guidebook for architects entitled 'Designing for the disabled'. In his book, Goldsmith tried to find answers to how architects can design buildings and facilities for people with mobility, visual, hearing and mental impairments. Published in 1963, Goldsmith's standard book of guidance

© Springer Nature Switzerland AG 2020
I. Shuayb, *Inclusive University Built Environments*,
https://doi.org/10.1007/978-3-030-35861-7_3

reflects the architect's perspective on, and rationale for, making separate and specific provision for individuals with disability. The book was recognised as a source of guidance for practising architects. It was used as a tool for designing special provisions for individuals with disabilities. Instead of reflecting the architect's own experience as an individual with disability, the book focused solely on specific provisions for wheelchair users, without acknowledging other types of disability.

Goldsmith was also involved in undertaking further research on the topic of designing for the disabled in 1963. The selection criteria were geared towards finding a large city with examples of all of the principal types of public building, and Norwich fitted these criteria. It was during his year of research that Goldsmith identified the most important need for individuals with disabilities, namely, a unisex public toilet. With internal dimensions of 1370 × 1750 mm, it was nearly twice as spacious as the toilet stall for wheelchair users prescribed in the 1961 American Standard. Although many CP96 toilet types were adopted in public buildings in the late 1970s, many wheelchair users complained that the space was not enough to allow a wheelchair to manoeuvre around. Further tests were conducted to enlarge the toilet space. In 1976 the 2000 × 1500 mm unisex toilet was adopted. Since then it has become a requirement of the Part M Building Regulations of the BS5810 code of practice issued in 1987 and 1992.

In 1967 and 1976 the second and third editions of the book were published. The three editions reflect the 1960s' and 1970s' paradigm that defined accessibility only as the removal of barriers for people with physical disabilities. The Chronically Sick and Disabled Persons Act was introduced in 1970. It called for the elimination of physical barriers to individuals with disabilities. Goldsmith was charged with implementing Section 3 of the 1970 legislation. His job was to provide new design guidance for special housing for wheelchair users.

In 1976, the British Building Regulations were published as a substantial package of technical provisions for building construction. Part T was the first legislative stage that enforced access requirements and elimination of physical barriers to individuals with disabilities in public buildings. Acknowledging that the 1976 regulation was complex and inflexible, a decision was made to recast and divide it into parts known as Approved Documents. Each document can be used as guidance to comply with the regulation requirements. It was proposed that Part T concerning access for individuals with disabilities should be drafted so it could be added to the new regulation. The revision of the 1976 regulation was introduced in 1985 and became known as the British Building Regulation which included 11 parts from A to L. Part T was based on BS5810, which specified the design standards to be used to ensure access for disabled people. There was no exit code provision for individuals with disabilities in the case of multistorey buildings; Part T compliance only required that single-storey public buildings should be accessible to individuals with disabilities. It is important to note that the normalisation and shift in attitudes towards an inclusive design that it was hoped would be achieved, firstly, as a result of the 1985 Part T, and secondly, the 1987 Approved Document M, failed to materialise.

In 1987 the Part M Building Regulation was introduced to enforce the requirements of the legislation, and developers and architects were required to take account of these regulations in their designs.

> The 1987 Part M Regulation was micro in form as it focused on catering only for individuals with disabilities. Moreover it only applied to new multistorey buildings that could be used as offices and shops. The number of lifts required for new multistorey buildings was revised in Part M and was determined by floor area (Goldsmith 1997, p. 98).

Although Goldsmith was responsible for setting the first building design standards in the UK, his micro-doctrine perspective, which considered individuals with disabilities to be different from able-bodied people, affected the way the first standards were written. Accordingly, the first Part M Regulation was called 'Access and facilities for disabled people' and required special and exclusive provisions and facilities for individuals with disabilities. Nugent, the originator of what is known now as 'the inclusive design' approach, believed that individuals with disabilities with specific needs can be accommodated by normal provisions.

Influenced by Nugent's inclusive approach, Goldsmith adopted the macro-doctrine approach and espoused a new paradigm. In 1997, he published his fourth edition of 'Designing for the disabled: The new paradigm'. This book reflects the architect's shift in thought towards the inclusive design paradigm. It includes relevant research findings and ideas based on the social model of disability and focused on the principles of inclusive design. In his findings, Goldsmith recognised that it is not only individuals with disabilities who may face physical barriers, but also women with infants and pushchairs. Through his research findings, he came to recognise that society and architectural barriers are the main reasons for disablement, and called for the adoption of the inclusive design approach in the revised Document M.

In 1988 the British Standards Institute issued BS5588 'Fire precautions in the design and construction of buildings. Code of practice for means of escape for individuals with disabilities'.

Part M Regulation was introduced in 1992. It aimed at providing access to people with physical, visual and hearing impairments; covered existing buildings, extensions to existing buildings and new constructions; and was applicable to all storeys in new and non-domestic buildings (Goldsmith 1997, p. 114).

Although the revised Building Regulations were improved to cater for other disabilities, Goldsmith (2001) and Imrie and Hall (2001) argue that they concentrate on designing for disability and do not accommodate the needs of other users, such as women and young children.

Both Goldsmith (2001) and Imrie and Hall (2001) state that Part M follows the top-down approach that only addresses limited types of disability. Goldsmith (2001, p. 4) argues that the exclusive attention paid to the needs of individuals with disabilities prevents the realisation of inclusive design as it ignores the wide spectrum of users who can be affected by the same barriers. An example of this is illustrated by the provision of public toilets for women, which is half the number of urinals and toilets provided for men.

Another limitation that Goldsmith (2001) highlights with regard to the Part M Regulation is that it just focuses on minimum design standards that can lead to the exclusion of many individuals with disabilities whose needs are not accommodated by these minimum provisions. Moreover, Goldsmith argues that Part M concentrates on designing for individuals with disabilities without including provisions for other users, which undermines designers' attempts to achieve inclusive design.

In addition, Imrie and Hall (2001) point out that the Part M Regulation is geared towards achieving and securing accessible environments for a limited range of disabilities and needs rather than legislating for the highest possible quality of accessibility. Whilst Part M Regulation included provisions for creating accessible features and facilities, it did not include provisions for people with visual and cognitive impairment, and those with dyslexia. The absence of provisions such as lighting, acoustics, information and directional signage, visual contrast and slip-resistant floor surfaces can be problematic for people with visual and cognitive impairments.

Whilst Part 3 of the DDA legislation embodies legal duties to eliminate physical barriers and enhances accessibility at buildings which are subject to Building Regulation compliance, Goldsmith (2001) and Imrie and Hall (2001) highlight a limitation in this requirement. They state that the lack of clarity in the wording has led architects and service providers to follow the minimum requirement of these accessibility guidelines stated in Approved Document M, which anticipate the needs of people with mobility and sensory impairments, and do not consider the various preferences and needs of a wider range of users.

According to Imrie and Hall (2001, p. 55), DDA legislation has placed duties on service providers and architects to remove the physical barriers and improve the accessibility at buildings without taking into consideration 'the broader infrastructure which characterises the built environment as a whole' Imrie and Hall (2001, p. 55). External pathways, pavements, street lighting, signage, material finishes and interior lighting were not included in the Approved Document M 1998, although they are all important features of the built environment that can affect the level of accessibility and for users with different needs and abilities. In acknowledgment of the limitations of Building Regulation Approved Document Part M and BS guidance, a shift in attitude towards the inclusive design approach is reflected in the new editions of BS 8300:2001 and its revised edition in 2009, as well as Part M 2004 and its 2010 edition. These editions embed provisions that tackle the needs of people with mobility impairment, people with visual and hearing impairments, children and elderly.

The inclusive design approach requires changes in attitudes and perceptions so as to understand the demands and needs of all potential users and not merely to acknowledge the needs of individuals with disabilities and of elderly people. To evaluate to what extent inclusive university environments have been achieved, this book investigates whether the Part M Building Regulations applied in six different university buildings over six decades, namely the 1960s, 1970s, 1980s, 1990s, 2000s and 2010s at the Universities of Essex, Bath and Kent, have impacted the design of buildings and achieved inclusive design. A detailed analysis of these findings is presented in Chap. 6. By examining the level of compliance with Approved

Document Part M, this book investigates whether such compliance has resulted in an appropriate response to the needs of a broad spectrum of users, including individuals with disabilities.

3.2 Inclusive Design

Whilst antidiscrimination legislation, such as the DDA in the UK, was introduced to prohibit discrimination against people with disabilities to gain access to education, employment and facilities and services, disabled activists and advocates of accessibility have put their efforts into eliminating physical barriers by proposing built environments that are accessible to people with disabilities. Acknowledging that accessibility concerns everyone, not just people with disabilities, Ron Mace, Ruth Lusher and other American accessibility advocates have urged the introduction of a different approach to the design of the built environment, which they call 'universal design' (Steinfeld and Maisel 2012, p. 27). This new approach aims to create environments, facilities and goods that respond to the diverse needs of users with different abilities and from different age groups. Ron Mace, the founder of the Center for Universal Design, defines universal design as:

> The design of products and environments to be usable by all people, to the greatest extent possible without the need for adaptation or specialized design (Story 2001, p. 10.3).

Universal design aims to improve products and environments for as many people as possible at little or no extra cost to benefit those with different abilities and from different age groups (Nussbaumer 2012, p. 29).

Various terms have emerged that share the same approach to designing for all users. The most common are 'barrier-free design', 'accessible design', 'lifespan design', 'designing for all', 'universal design' and 'inclusive design'. Whilst universal design is mainly used in the USA, inclusive design is another term that is used in the UK, Canada and other European countries.

3.2.1 Inclusive Design Definition

The term 'inclusive design' originated in the UK and refers to 'the provision of quality of life and independent living for the ageing population' (Waller and Clarkson 2009, p. 19). In Britain, the most commonly used definition for inclusive design is

> The design of mainstream products and/or services that are accessible to, and usable by, as many people as reasonably possible on a global basis, in a wide variety of situations and to the greatest extent possible without the need for special adaptation or specialised design (British Standards Institute 2005).

The Commission for Architecture and the Built Environment (CABE) defines inclusive design as the process of making places that everyone can use (CABE 2008). The intent of inclusive design aims to improve the quality of life for potential users by creating environments, spaces, goods, gadgets and means of communication that can be used by a variety of users regardless of their particular need or ability at affordable cost. Hence, inclusive design is beneficial for many users with different abilities and ages (Clarkson et al. 2003).

Inclusive design goes further than designing for people with disabilities as it involves producing designs, environments and goods that everyone can use. According to CABE (2008), inclusive design aims to eliminate physical barriers that promote segregation and require users to put extra physical effort. It allows users with different abilities to 'function equally, confidently and independently in the course of their everyday activities. An inclusive approach to design offers new insights into the way people interact with the built environment. It creates new opportunities to deploy creative and problem-solving skills' (CABE 2008, p. 3).

There are many different terms that are used in describing inclusive design. Some scholars employ the terms 'universal design' or 'barrier-free design'. Many terms have similar and overlapping definitions; however, they all consider the needs of as many people as possible including parents with children, elderly people and disabled individuals.

Mace (1985) defines universal design 'as the design of products and environments to be usable by all people, to the greatest extent without the need for adaptation or specialised design' (Story 2001, p. 4). Steinfeld and Maisel (2012) and Imrie and Hall (2001) argue that parts of Mace's definition and its terms, such as 'all people', 'greatest extent' and 'without the need for adaptation or specialised design' (pp. 17–18), can be misinterpreted by professionals. These terms do not completely embrace the inclusive design purpose as they can lead professionals to avoid making provision for individuals with disabilities.

Acknowledging that universal design/inclusive design aims to promote social inclusion, equality and independence, Steinfeld and Maisel (2012, p. 29) define universal design as 'a process that enables and empowers a diverse population by improving human performance, health and wellness and social participation'.

Whilst 'universal design' is mainly used in the USA and 'inclusive design' is used in the UK and many European countries to describe the process of designing for all users, both terms differ from 'accessible design'. Accessible design aims to create products and environments that are especially suitable for individuals with disabilities (Nussbaumer 2012, p. 28). Inclusive design aims to involve users in the design process to develop products and environments that can be affordable and used by people with diverse needs and different abilities and ages, in order to remove the negative social and attitudinal barriers that are associated with the building and design market (Ormerod et al. 2002).

Goldsmith (2001) defines universal design as a design approach that accommodates and caters for a wide range of users and emphasises the necessity to implement the bottom-up route to universal design. To illustrate that, he gives an example of a universal design product, namely automatic doors that are provided in main entrances of building and can be used by users with different abilities. Goldsmith stresses that an architect who takes the bottom-up route to universal design takes into consideration the different users' needs and ages, and therefore caters for people with disabilities.

To promote the adoption of universal design, the Center for Universal Design (CUD) at the College of Design, North Carolina State University, has developed seven principles of universal design that allow an accessible environment for all to be achieved. Table 3.1 presents these seven principles.

Steinfeld and Maisel (2012, p. 87) point out that although the seven principles of universal design are a valuable attempt to describe the scope of universal design and provide guidelines for practice, the lack of clarity in some of the principles has led to their being misinterpreted. One of the main criticisms is that many of these principles can be more suited to product design rather than other design disciplines. Another criticism that Stenfield and Maisel (2012, p. 88) noted is that many of these principles may be misinterpreted or misunderstood since they 'lack clarity, and some principles overlap in terms of their objectives and goals'. Thus, in order to have a better understanding of universal design and achieve its widespread adoption, Steinfeld and Maisel (2012) have developed eight goals of universal design that aim to improve human performance, health and social participation. The first goal 'body fit' acknowledges the people's wide range of body sizes and abilities by creating designs that accommodate different body sizes and hence meet Stenfield and Maisel's (2012, p 90) second and third goals 'comfort' and 'wellness' to guarantee that designs are safe, secure and comfortable. Moreover 'awareness and understanding' are the third and fourth goals that they proposed to ensure that the universally designed products and spaces provide clear information for the people to use and operate. Moreover, 'social integration, personalisation and cultural appropriateness' are goals that stress out the importance of choice, preferences, equal opportunities and respect for people's cultural background and values.

Table 3.1 The seven principles of universal design

Principles	Descriptions
1. Equitable use	The design is useful and marketable to people with diverse abilities.
2. Flexibility of use	The design accommodates a wide range of individual preferences and abilities.
3. Simple and intuitive use	Use of the design is easy to understand, regardless of the user's experience, knowledge, language skills or current concentration level.
4. Perceptible information	The design communicates necessary information effectively to the user, regardless of ambient conditions or the user's sensory abilities.
5. Tolerance of error	The design minimises hazards and adverse consequences of accidental or unintended actions.
6. Low physical effort	The design can be used efficiently and comfortably and with a minimum of effort.
7. Size and spaces for approach and use	Appropriate size and space are provided for approach, reach, manipulation and use, regardless of user's body size, posture or mobility.

Center for Universal Design (CUD 2008)

Although universal design and inclusive design aim to eliminate physical barriers and create accessible built environments, 'the universal design approach does not attempt to reconcile the often conflicting needs of every possible minority group in society, while inclusive design proposes many design solutions that tackle different needs to break down unnecessary barriers and exclusiveness' (Imrie and Hall 2001, p. 18). According to CABE (2008, p. 3), inclusive design aims to promote social participation and integration by 'enabling everyone to participate equally, confidently and independently in everyday activities'. CABE (2008) states that creating of an inclusive environment can be achieved by the collaboration of end users with a variety of professionals in charge of creating built environments and spaces. Collaboration between end users and professionals such as architects, civil and electrical engineers, and interior and landscape designers, in addition to access consultants, can lead to creating a built environment and architecture spaces that respond to the diverse users' needs. In addition, to ensure that created spaces and environments are managed inclusively, landlords, service providers and staff members should also be involved. Thus inclusive design involves more than designing for people with disabilities. It is a holistic approach that roots itself in the participatory approach by including everyone in planning, designing and managing spaces and buildings, whereas universal design 'considers the global aspect, occurring everywhere and available to everyone' (Nussbaumer 2012, p. 34). According to Imrie and Hall (2001, p. 18), 'inclusive design has much in common with Sommer's (1983) conception of a social design approach that places users at the heart of the design process rather than at its margins'. An inclusive design approach based on participation empowers the users, giving them the right to take control over their environments, and the chance to communicate their particular needs to design professionals. The collaboration of users and professionals will enable architects and designers to interlink their design skills effectively with the experiential knowledge of users to achieve an inclusive environment for people with different abilities and from different age groups (Imrie and Hall 2001; CABE 2008).

3.2.2   Ethos of Inclusive Design

There is an increasing demand in adopting inclusive design among many countries due to the increase in the number of older population and people with disabilities. Moreover, the legislation is focused on applying normal provision that will benefit the whole population. With the notion of globalisation, and the increased number of population with different cultural backgrounds, ethnicities, abilities and needs, creating inclusive spaces, products and facilities is essential in responding to the diverse people's needs. People within the population have a range of different capabilities and needs which have influenced design guidelines so that they focus on a variety of needs instead of on the needs of the notional 'average' person.

According to Sawyer and Bright (2007, p. 1), 'design guidance is always based on the needs of the so- called "average" person; however, everyone differs from the average in some way. People vary in terms of height, strength and dexterity. People have different visual and hearing abilities or may have respiratory impairments or reduced stamina'.

According to recent demographic studies in the UK, the population is shifting towards becoming an increasingly elderly one. By 2020 half of the UK adult population is expected to be over 50, which means that the built environment has to cater for such a group. Steinfeld and Maisel (2012) argue that the demographic shift towards an ageing population and the social movement towards integrating individuals with disabilities into the mainstream society provide the foundation for the adoption of inclusive design. These factors have resulted in new markets and businesses which have developed products, environments and gadgets that can be used by a diverse range of users from different age groups, and with different abilities.

An inclusive design approach does not only benefit individuals with disabilities; there are also around 18 million people in the UK who would directly or indirectly benefit from inclusive access to buildings and public spaces. These include older people, families with children under the age of 5, carers and the friends and relatives who accompany people with disabilities (Department for Communities and Local Government 2003; Goldsmith 2001).

Goldsmith's inclusive design pyramid (2001) reveals that architects following the bottom-up route to inclusive design propose design solutions that can be used by different users including those with disabilities, since they consider them as normal people. He explains that an architect following the bottom-up approach is taking into consideration the different users' needs when moving from a row to another, in order to propose inclusive designs that anticipate the diverse needs, and avoid creating special designs that promote segregation and isolation for people with disabilities. An important finding from Goldsmith's inclusive design pyramid reveals that rows 3, 4 and 5, which contain able-bodied people, such as women, children and old people, using public toilets in public buildings, are subject to architectural discrimination because normal provision is not suitable for them. Such a finding highlights the importance of adopting the inclusive design approach so that architects will design buildings that provide adequate numbers of public toilets to accommodate the needs of women, children and people with disabilities, as well as steps and stairs with nosings and equipped with handrails on both sides.

Many architects start with an assumption that people with disabilities are different and do not fall within the average category, which results in their needs being ignored when designing buildings, whilst in other cases there is provision of specially designed solutions. This approach is called the top-down approach to design: considering individuals with disabilities as different promotes segregation from the mainstream society. Steinfeld and Danford (1998) point out that inclusive design eliminates the use of 'us' and 'them' distinctions. The inclusive design paradigm promotes equality and diversity by offering built environments and facilities that can be used by children, teenagers, families, women, elderly and people with different abilities and cultural and ethnic backgrounds. Hence it surpasses accessible design that is geared towards people with disabilities' needs.

Influenced by the inclusive design approach, Approved Document Part M 1995 was amended and Part M 2004 was introduced to include such provisions in the standards. Whilst the inclusive design approach aims to include the needs of all users, Sawyer and Bright (2007) emphasise that inclusive design also offers special designs and assistive auxiliary devices for people who may prefer to use such special provisions. Offering a choice for users is one of the main principles of inclusive design. Thus by providing a ramp along a staircase; a textbook, Braille or audio book; hearing enhancement system; or a screen with subtitle, people are offered to choose their preferred provisions that respond to their particular need.

Since the Disability Discrimination Act (DDA) was introduced in 1995, service providers have had to make reasonable adjustments to provide access for disabled people. 'To ensure that there is reasonable provision of access to all users, access statements were introduced in the Planning and Compulsory Purchase Act 2004 and took effect from August 2006' (Sawyer and Bright 2007, p. 105). Access statements have been introduced as part of the planning and building application submitted for approval, to ensure that accessibility provisions are considered when constructing a new building or adding an extension to a listed heritage or existing building (Building Control Guidance Note 2006).

> An access statement provides an opportunity for developers, designers and managers of buildings and environments to demonstrate their commitment to accessibility and to set out and record issues relating to accessibility through strategic design, construction and occupation of a project. In the initial stages of a project an access statement can be used to record the elements of the brief that relate to access. At this strategic level it will be a statement of intent and can demonstrate how the project will meet any relevant legislation (Sawyer and Bright 2007, p. 35).

An access statement is introduced as a way of raising awareness about inclusive design. It is considered a useful method that enables architects and designers to propose inclusive provisions for new buildings and extensions. Moreover, an access statement allows architects and designers to ensure that their design proposals comply with the Building Regulation such as Approved Document M, or BS 8300.

An architect or a designer is required to include information about the accessibility provisions and procedures that he/she proposes to ensure that the design proposal is accessible for people with different abilities and needs. Hence the access statement should include:

- A brief explanation of the client's policies and approach to access
- Sources of advice and guidance on access which will be followed
- Details of any consultations undertaken or planned
- An explanation of any specific issues affecting accessibility and details of access solutions proposed, including those which deviate from recognised sources of good practice
- Details of the management and maintenance management policies adopted, or to be adopted, to maintain features enhancing accessibility (Sawyer and Bright (2007, p. 35)

Sawyer and Bright (2007, p. 35) point out that where good accessibility practice cannot be achieved, the access statement should provide the reasons, explain the implications for the users and propose other solutions to lessen any adverse effects on accessibility. Moreover, the access statement has to include a justification for any proposed design feature which provides an equivalent standard of accessibility if the designer decides not to follow the current design guidance. A participatory approach and consultations with end users, including individuals with disabilities, are useful methods that can be used during the design process to draft an access statement that takes into consideration the diverse consulted users' needs.

The access statement progresses into a document with the development of the project. Although it starts at the strategic level, the access statement will include detailed information which describes the proposed accessibility provisions and design suggestions along with the management practices and procedures that are used during the design and implementation phases of a project. According to the Building Control Guidance (2016), this information is always revised and checked during the design and the implementation phases, or when a building undergoes alterations that may affect the level of accessibility for people who use it.

During the planning phase, the access statement should provide information about the site plan and means of transportation that serves the site including car parking bays. Moreover, the access statement should state the design guidelines used to meet the legislation requirement concerning accessibility and information that addresses the emergency exits and means of escape used (Sawyer and Bright 2007).

The access statement document includes more detail about technical issues and describes the design decisions which will be submitted with a building regulation application. This document is considered an evidence-based defence document that can be used in court in the event of any legal challenge to justify actions taken (Building Control Guidance 2016). An example of an inclusive access statement for the Jarman Building at the University of Kent (see Appendix A) describes the inclusive design approach adopted. It covers the accessibility barriers in the approach to the site and public transport, entrances and lifts, as well as circulation, stairs, toilets and showers, signage and navigation, and lighting and decor. It is important to note that Chap. 6 includes a detailed discussion of the Jarman Access Statement and whether it was used in the detailed design and construction phases.

Although the accessibility criteria relate to the disability legislation and the application of Approved Document Part M and other access design guidance, such as the BS 8300, in the UK, which require accessible environments to be provided, inclusive design goes beyond the minimum requirements set by the accessibility standards and aims to consider the needs of all users, regardless of age group, ability and size. It is economical, attractive and marketable, as it offers choice, flexibility and diversity so that people of various ages and abilities can use the building or space. A misconception held by many architects, planners and contractors is that inclusive design adds cost. However, such an argument has been challenged by inclusive design projects, which have revealed that additional costs were minor if inclusive design was adopted during the design process and applied throughout the implementation. Sawyer and Bright (2007, p. 2) note, 'careful consideration of accessibility issues at the design stage and good management throughout the life of a building can offer and sustain accessible environments at little or no extra cost'. Maintenance and refurbishment plans can be used as opportunities to resolve accessibility barriers and achieve inclusive environments with minimal or little additional expenditure.

According to Imrie and Hall (2001 p. 41), the Habinteg Housing Association has estimated that lifetime homes that were built in 1999 'cost £300 to £400 more to build than conventional housing stock'. Lifetime homes, as illustrated in Fig. 3.1, are designed to comply with 16 design criteria based on inclusive design. Moreover,

Lifetime Homes Diagram

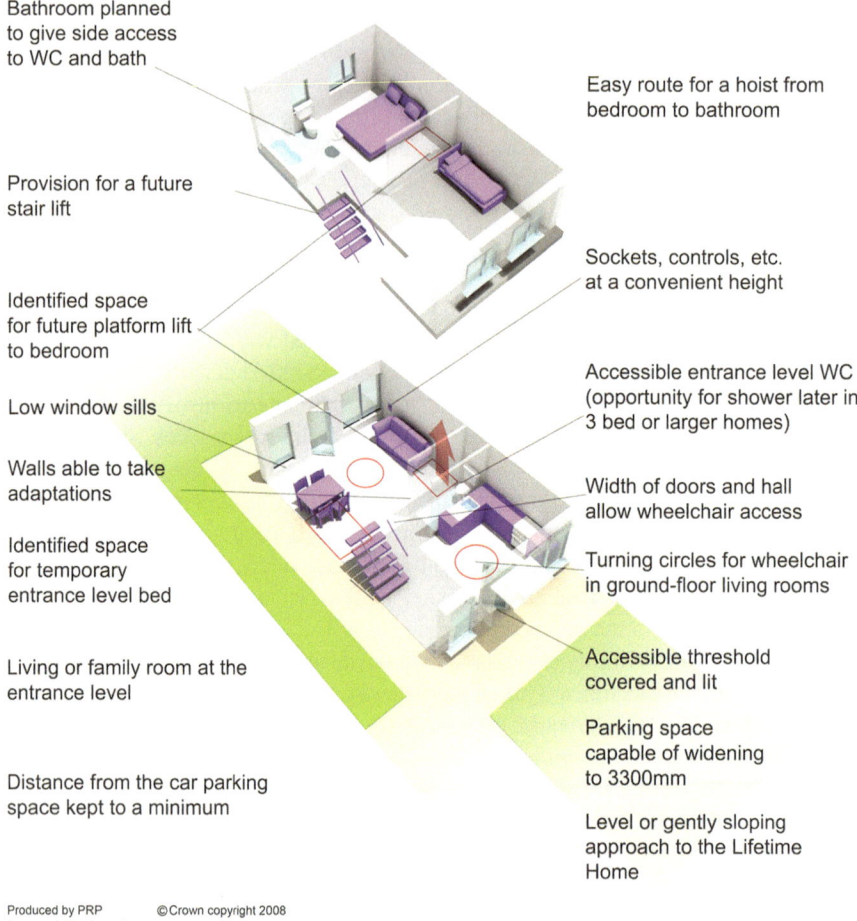

Bathroom planned to give side access to WC and bath

Easy route for a hoist from bedroom to bathroom

Provision for a future stair lift

Sockets, controls, etc. at a convenient height

Identified space for future platform lift to bedroom

Accessible entrance level WC (opportunity for shower later in 3 bed or larger homes)

Low window sills

Walls able to take adaptations

Width of doors and hall allow wheelchair access

Identified space for temporary entrance level bed

Turning circles for wheelchair in ground-floor living rooms

Living or family room at the entrance level

Accessible threshold covered and lit

Parking space capable of widening to 3300mm

Distance from the car parking space kept to a minimum

Level or gently sloping approach to the Lifetime Home

Produced by PRP   © Crown copyright 2008
Diagram indicative only

Fig. 3.1 Lifetime homes diagram of 12 provisions. 1- Level or gently sloping approach to the home with accessible threshold covered and lit. 2- Parking space wider than 3300mm. 3- Turning circles for wheelchair in ground floor living rooms. 4- Wide doors and hall for wheelchair access. 5- Accessible entrance level wc and bath to give side access on ground floor . 6- Easy route for a hoist from bedroom to bathroom. 7- Living or family room at the entrance level. 8- identified space for temporary entrance level bed. 9- low window sills. 10- identified space for future platform lift to bedroom on 2nd floor. 11- Walls able to take adaptations. 12- Provision for a future stair lift

the Access Committee for England (1995, p. 10) estimated that the average cost of enhancing accessibility varies between £180 and £400, which is considered to be a small price if compared to the long-term benefits that can be achieved by providing an inclusive and accessible building or space. Imrie and Hall (2001, p. 41) point out, 'the Department for Education and Employment has conducted a survey of 700 firms in England and Wales to estimate the costs of making "reasonable adjustments". Findings from this survey revealed that 44 per cent of adjustments cost less than £49 and only 5 per cent were more than £5000'. Levine (2003, p. 10), an architect and Assistant Director of the Inclusive Design and Environmental Access (IDEA), points out that buildings designed inclusively from the beginning to be usable by a wide range of users will need fewer alterations/refurbishments in the future. Steinfeld (2010) notes that truly accessible and inclusive features added to a home have the potential to add value to it. Steinfeld and Maisel (2012, p. 46) state that the demographic data on household expenditure can identify trends towards new spending patterns that inclusive design can address with regard to the USA. They believe that inclusive design can be applied to public transportation in order to provide pleasant and convenient means of transport that can be used by a wide range of users to resolve the problem of a rising percentage of household income being devoted to transport due to increased energy costs. Moreover, demographic factors play an important role when it comes to providing inclusive environments and products that people of different age groups can benefit from. Steinfeld and Maisel (2012) state, 'older households have substantially more assets than younger ones and despite reduced income, elderly people have more money to spend because their expenses are lower' (Steinfeld and Maisel 2012, p. 46).

3.2.3   Principles of Inclusive Design

A review of the literature shows that there are five principles of inclusive design developed by the Chartered Association of Building Engineers (CABE). This section presents and compares CABE's inclusive design principles with the seven principles of universal design. According to CABE (2008), the first inclusive design principle 'places people at the heart of the design process' by involving users in the design process and ensuring that the design responds to the users' diverse needs. It values the users' 'diversity and difference' which is CABE's second principle, by creating environments and services that consider the different users' ages, abilities and cultural backgrounds, and hence provides 'choice' which is the third inclusive design principle. Moreover, 'flexibility of use' and 'convenient and enjoyable for all' are two essential principles in inclusive design as they aim to provide designs that adapt to the diverse users' needs and demands and allow them to equally enjoy using the inclusive designs and facilities.

Whilst the first inclusive design principle places 'people at the heart of the design process' by involving a wide variety of those with different abilities in the design process, universal design principles do not involve end users in the design process.

Anticipating the needs of a broad spectrum of users without involving them in the design and implementation phase does not necessarily lead to achieving inclusive environments or products that all users can benefit from. Involving users, contractors, surveyors, architects and designers during the design and implementation phases will promote social integration, the aim of which is to treat all people with dignity and respect, which in turn is considered to be one of the eight goals of universal design that Steinfeld and Maisel (2012) propose.

The second principle of inclusive design, which acknowledges the 'diversity and difference' of users, is on a par with the 'equitable use' principle of universal design. By identifying barriers that can impede a wide range of users with different abilities and from different age groups, and by anticipating their diverse needs and abilities, inclusive design can achieve two of Steinfeld and Maisel's (2012, p. 90) goals of universal design, namely the 'body fit' and 'cultural appropriateness'.

'Choice' is the third principle of inclusive design which is related to the 'simple and intuitive use and low physical effort' (Steinfeld and Maisel 2012, p. 90) principles of universal design. Although universal design suggests that one single design solution can accommodate the requirement of all users, inclusive design highlights the need to offer different design solutions to respond to the diverse needs and abilities of users. Offering choices that are simple and easy to use by people with different abilities and from different age groups will respond to the 'body fit, comfort, understanding, wellness, and personalisation' goals of universal design (Steinfeld and Maisel 2012, p. 90).

Universal and inclusive design principles share the same goal of 'flexibility of use' which aims to create an environment or a product that adapts to changing uses and demands. This principle responds to the 'body fit and comfort' goals of universal design suggested by Steinfeld and Maisel (2012, p. 90). In addition, providing environments and products that are 'convenient and enjoyable to use for everyone' is one of the main principles of inclusive design. This principle coincides with 'perceptible information, tolerance for error, and size and space for approach and use' principles of universal design (Steinfeld and Maisel 2012, p. 90).

By offering accessible transportation means, roads, pedestrian routes, internal entrances, signage and virtual information that minimise hazards and adverse consequences of accidental or unintended actions, inclusive design responds to the eight goals of universal design that Steinfeld and Maisel (2012) propose.

Nussbaumer (2012) has also used these principles to propose a set of inclusive design criteria. For an environment, service or product to be inclusive, and to fit within Nussbaumer's inclusive design criteria, it should be 'responsive, adaptable, accessible, and secure' (p. 37). Hence it responds to the diverse users' needs and abilities, adapts to the diverse needs of its users, complies with accessibility guidelines and building regulations, and provides safety and physical and psychological comfort for its users.

Whilst the comparison between inclusive and universal design principles highlights similarities and differences between the approaches, Steinfeld and Maisel's (2012) goals of universal design incorporate both universal and inclusive principles. Both Steinfeld and Maisel's (2012) goals and Nussbaumer's (2012) inclusive design criteria will be used in this study to investigate whether the universities, which are the subjects of this research, responded to these criteria when making their built environment accessible to potential users.

3.2.4 Inclusive/Universal Design in Professional Educational Programmes

In the USA, since the introduction of the Rehabilitation Act 1973, efforts have been made to eliminate physical barriers and hence to provide access for disabled people. To ensure that architects would become more aware of the needs of individuals with disabilities, a new programme was introduced at the University of California Berkeley in 1979 by Raymond Lifchez, a professional educator, who conducted an experimental project known as 'Architectural design with the physically disabled user in mind' (Lifchez 1987). This programme aimed at testing the participatory approach and methods that would enable students to develop the skills needed to bridge the gap between the needs of able-bodied people and those of individuals with disabilities. The collaboration with users with physical impairments enabled students to anticipate the users' needs and create environments that would address their impairments.

In the UK in 1963, Selwyn Goldsmith was the first architect to introduce accessibility in a guidebook entitled 'Designing for the disabled'. Goldsmith, infected by a polio virus, became disabled at the age of 23. In his book, he tried to discover how architects could design for people with physical impairments.

Since the revision of universal design, that was led by Ron Mace in 1985, educators with an interest in the subject have initiated several projects to improve universal design education in design schools. Elaine Ostroff was the first educator in the USA to lead the Universal Design Education Project (UDEP), sponsored by the National Endowment for the Arts (NEA) (Steinfeld and Maisel 2012, p. 75). The UDEP project inspired other European educators to integrate universal design into their educational curriculum.

In the UK, inclusive design has been taught and researched for some years using a variety of methods across built environment disciplines; however, its mainstream adoption in the built environment curricula remains rather limited (Inalhan 2012). The Architectural Association School in London offers a graduate diploma in Environmental Access, whilst the University of Reading offers an MSc in Inclusive Environments. The University of Reading programme is cross-professional and is aimed at those involved in the design and management of built environments. The University of the West of England offers a BA in Architecture and Planning which embeds inclusive design throughout the course, whilst the University of Sheffield introduces inclusive design as one of the guiding principles of the design studio course for first-year architecture students. The University of Salford offers a distance-taught, multidisciplinary online programme for professionals which is oriented towards accessibility and inclusive design. These programmes include a postgraduate certificate, postgraduate diploma and Master of Science (Marrow et al. 2003). The Helen Hamlyn Research Centre at the Royal College of Art has an Age and Ability Laboratory that focuses on and offers a graduate degree, in inclusive design (Steinfeld and Maisel 2012, p. 75). Inalhan (2012, p 29) states that many of these programmes only address inclusive and universal design in terms of complying with accessibility standards regarding disabled people's needs, which leads architectural students to underestimate the importance of this learning and to fail to absorb the information provided since it mainly focuses on disability needs rather than the needs of a wide range of users.

Moreover, Inalhan (2012) points out that the inability of students to transfer the knowledge gained from these courses to architectural project work in practice results in their failure to apply the inclusive design principles. According to Inalhan (2012, p. 30) students' and architects' lack of awareness of the ethos of inclusive design and their inability to recognise the population's diverse needs or consult with, and listen to, users lead them to fail to achieve inclusive design environments. Steinfeld and Maisel (2012, p. 74) point out that an important factor involved in increasing the rate of inclusive/universal design adoption is the need to integrate inclusive/universal design into professional education and continuing education programmes.

In acknowledgment of the importance of professional education in raising awareness of inclusive design and preparing architectural graduates to adopt inclusive design in their designs, Chap. 5 of this book examines architects' knowledge about inclusive design and whether they consult with end users during the design and implementation phases. Moreover, Chap. 5 highlights the training and professional courses that architects have undertaken to expand their knowledge about accessibility and inclusive design.

3.2.5 Inclusive Design Case Studies

Whilst some academic researchers, such as Goldsmith (2001), Imrie and Hall (2001), Clarkson et al. (2003), Burton and Mitchell (2007) and Bright and Cook (2010), have based their work on assessing products, services and built environments, there is little literature about the impact of inclusive design on architectural buildings and built environment. Accordingly, and in order to understand the inclusive design approach and its position in relation to eliminating architectural barriers and promoting inclusion, a review of different research projects adopting the inclusive design approach is presented in this section. The University of Arizona represents a case study that has used geospatial techniques to evaluate the level of accessibility. The Carrington Building at the University of Reading, the Willows School in Wolverhampton, the Eden Project gardens and the Roundhouse are four successful inclusive design case studies that illustrate the principles.

3.2.5.1 Universal Design and Accessible Space at the University of Arizona

The Disability Research Center at the University of Arizona involved students with disabilities in a project to evaluate the accessibility of the campus in 2008. The project used qualitative and geospatial techniques with the aim of engaging 15 students with and without disabilities and 5 staff members in taking part in 'map interviews'. The interviews solicited information about the paths that people used, their interactions with laboratories and classrooms, and the perceptions and attitudes about disability that campus users encountered. In each interview, participants were shown maps of the campus and were requested to use highlight and draw on a paper-sized map the passageways and routes they use on campus. A colour coding system was used, where blue colour indicated the passageways and routes used for daily travel. Red colour was used to point out inaccessible spaces and areas that promoted segregation and isolation. Green colour highlighted the accessible spaces which acknowledge the diverse needs of users. Moreover, personal interviews were conducted to obtain information from users about experiences in using the facilities on campus and the physical and attitudinal barriers that they face, and their vision and perception of an accessible campus in future. The findings from this project revealed that the University of Arizona campus did not fully anticipate the needs of all users. Feedback from users revealed that the level of accessibility varied from one building to another. Whilst the new Poetry Center was considered to be a creative and inclusive new building by some users, one interviewee with visual impairment considered it the 'worst' as it was constructed primarily from glass. The interviewee stated that lighting conditions inside auditoriums created shadows and glare which restricted many students with visual impairments from taking notes and accessing the taught information. Such a finding demonstrates that the way in which the building was designed did not anticipate the needs of people with visual impairments and other users who were impacted by the same barriers. Another finding from this research highlighted that both students with and without disabilities had to change their strategies to adapt to the University of Arizona campus. One of the main adaptive strategies that many users had to adopt was walking, cycling or driving along different routes on campus at night to avoid poorly lit areas. Respondents who lived within walking distance stated that although the campus did not have enough parking spaces, they chose to drive at a certain time of day as they had fears about tripping on kerbs that were less visible in the dark, as well as personal safety concerns. Another important finding from this research was that both physical and attitudinal barriers existed at the University of Arizona campus. Many interviewees and student researchers pointed out that the lack of awareness about different disability needs and the low financial resource allocation had limited the adoption of universal design on campus. Some participants pointed out that the transportation services, such as the bus shuttle, were not accessible to people with visual impairments. Another interviewee pointed out that such barriers should have been removed a long time ago as there have been advocacy and awareness campaigns on campus about these barriers. This research highlights the importance of improving both the physical and management facilities at the University of Arizona in order to achieve an inclusive environment (Rattray et al. 2008).

3.2.5.2   The Carrington Building, University of Reading

The Carrington Building at the University of Reading adopted the inclusive design principles and ethos, and is an example of how the built environment can tackle accessibility and environmental issues. This resulted in winning the 2010 Office Depot Green Innovation Award. To achieve inclusion and reduce energy consumption and CO_2 emissions, the University encouraged users to generate their own ideas and contribute to the project.

The Carrington Building accommodates the following facilities: Disability Office, Careers Advisory Service, Centre for Career Management Skills, study advisors, Counselling Service and IT help desk. The three-storey building incorporates a unique feature in its external façade, using recessed coloured stripes within the render, which correspond to the colour coding system used to distinguish the three floor levels (Bright and Cook 2010). Moreover, the building offers flexibility in terms of access to the main entrance by providing level and stepped entrances with handrails running along the staircase. The layout, position and colour scheme, lime green and mahogany for the reception area, were carefully designed so that it caters for all potential users. A low-level reception counter has been provided so wheelchair users can manoeuvre around, clearly signed induction loops have been installed so people with hearing impairment can communicate with the receptionist, and adequate lighting is suitably positioned so it does not create glare or shadows for lip-readers. In addition, the reception help desk is clearly signposted with adequately sized text that contrasts well with the plain rear wall behind the desk, allowing people with visual impairment to identify the reception counter and permitting them to see the faces of the staff.

In order to assist people to navigate around the building and understand the facilities provided on each floor, a colour coding system has been adopted inside the building where the door frames, architraves and skirting have different colours that correspond to the external colour features provided in the rendered façade.

Internal stairs have been carefully designed so steps are clearly visible with contrasting nosings, and there are handrails running along steps and landings. Moreover, the common seating areas provide flexibility for people to use either fixed or movable seats that are clearly visible against the surrounding surfaces. The toilet compartments that are provided include wheelchair-accessible toilets; they are also carefully designed to cater for people with visual impairments by providing adequate colour contrast between sanitary units, grab rails and surrounding surfaces.

3.2.5.3   The Willows School in Wolverhampton

The Willows School in Wolverhampton aimed at creating one building campus that includes a special primary school, and community facilities. Green Park Special School accommodated 100 students who have different physical, cognitive, intellectual and sensory impairments aged between 4 and 19 years old. Hence, there was a necessity to enhance accessibility for students with different needs and abilities.

Furthermore, there was a need for a facility to provide educational development for staff and faculty members, improve the student's educational and extracurriculum activists and deliver training sessions for relatives, nurses and carers. On the other hand, Stowlawn School was a primary school which had two different buildings serving students and staff members. The main barrier noted by Stowlawn School's staff members was that its two buildings promoted segregation and isolation and thus there was a necessity to design a new school that takes into consideration the staff's and students' different needs.

The Willows School was designed by Architype architects who were keen to design the school so it would be ecological, green and sustainable. Architype, adopting the inclusive design approach, collaborated and worked with users, students, staff members and teachers, and carried out a great deal of design discussion to achieve a comfortable, inspiring and functional space. Architype's understanding of inclusive design is expressed in one of its architect's statements:

> The inclusive design is not just about meeting the technical requirement of access for everybody whatever their abilities are, it is about including people in the process of the design so they become part of the building. What architects build using an inclusive process is generally a better building rooted in the ideas and dreams and aspirations of the people who will use the building … Without involving them we would not know how things function (Royal Institute of British Architects 2009a).

Aiming to combine special needs schools, a primary school and a community element in one project, architects went beyond simply accommodating a structure for a hoist for disabled children and made it a fundamental part of the design so no one would particularly notice the difference between going to classes in the primary school and going to those in the special school.

It is important to note that consultations with users drew the attention of architects to tackling the needs of children with autism who had problems with angular space. Accordingly, curvilinear flowing spaces became part of the design strategy in the building, and thus such curvilinear spaces were provided throughout the reception and common areas and space to manoeuvre was made available for people with mobility and visual impairments whilst simultaneously tackling the needs of autistic children.

Acknowledging the demand in special schools to maintain a slightly higher temperature in some areas, the architecture team looked in great detail, in collaboration with electrical and heating engineers, at how to provide a satisfactory level of ventilation by using heat exchange technology so that a good flow of air could be achieved, in addition to using cross-ventilation, which could be produced by having the windows open. This approach was greatly appreciated by the head teacher of Green Park School, who noted:

> They got rooms that are well ventilated and are very warm. I think that flexibility and adjustment across the campus is very important (Royal Institute for British Architects 2009a).

Daylight was also carefully considered, and the architects decided to use triple-glazed windows that generated a general level of good daylight with glare being controlled. In addition, the flexibility in the design approach allowed the provision of cupboards of different heights, which could be easily reached by children and teachers.

Because many staff members highlighted a major issue they faced with regard to using staff toilets, which were located in the old school building far from classrooms and facilities, the new inclusive design strategically placed the new toilets so everyone would have equal access.

As part of the consultation, children were involved in examining and appraising the design, and were allowed to see and touch conventional models and 3D computer simulations. Moreover, the contractor engaged with them during the construction phase, allowing them to see the building with their own eyes.

3.2.5.4   Eden Project Gardens

The Eden Project gardens aimed to create an accessible and inclusive environment for all its users, and worked with the Sensory Trust, which is a national organisation in the UK that promotes and supports inclusive design. The Sensory Trust works to provide easy access to spaces and remove architectural barriers, not only for individuals with disabilities and the elderly, but also for young children and mothers with pushchairs, who are equally impacted by the same barriers. The director of the Sensory Trust said:

> If you think about somebody with young children and mothers with pushchairs and family and friends, they are equally impacted by the same barriers. As a designer designing a space, I can't only think about an average person, I have to think of people in terms of their primary sense, maybe they don't have sight or hearing, and what we know from experiences, what comes up as an end product, when people have really taken it on board, is a design that is way more interesting for people in general (Royal Institute of British Architects 2009b).

The Eden Project gardens, being an inclusive environment, provide ramps and routes that make it easy for people to get around, non-slippery surfaces that are level, easy access for wheelchairs and pushchairs, accessible toilets and tactile warnings that alert partially sighted people to hazards, such as approaching stairs.

According to the Sensory Trust Director, legislation, such as DDA Part M, has helped architects to enhance accessibility, but she believes that if architects just stick to the standards, there is a danger that the design proposed will treat individuals with disabilities as a specific group of people, instead of including them as part of the crowd, regardless of their age or their abilities. To achieve an inclusive environment, the Director of the Sensory Trust emphasised the importance of involving people in consultation and acknowledging their different needs and experiences, so that architects or designers will interpret these views through their designs and will not rely on disabled people to design the space on their behalf.

> I find that consultation can be a quite a threatening thing for designers, and this comes from a slight misconception of what consultation needs to be. It is very easy for a designer to think that you get a lot of people in and ask them to design something on your behalf. In our view, the design skills sit with the designer, the consultation is about bringing into play the different views that people have, the different ideas they have, the different experiences that

they want to have, and really what that does is that it brings in a whole load of things that the designers have to take on board and interpret through their designs (Royal Institute of British Architects 2009b).

Moreover, a landscape architect at the Eden Project gardens highlighted the importance of consultation in balancing people's different needs.

To make a design inclusive, it is a really a challenge and it can probably seem overwhelming to try and think of all the things you need to think of. But again, talking to people brings out the key things that you need to include and also helps you to solve conflicts, because sometimes things that work well for one group or one particular range of disabilities work against another group so there is a balance to be reached (Royal Institute of British Architects 2009b).

Accordingly, consultation was an important method used in the Eden Project gardens, with individuals with disabilities having the chance to consider different materials, plants and textures, and comment on their experiences. Such feedback played an important role in providing new features in the Eden Project gardens, such as a raised platform that wheelchair users, young children and short people could use to overlook the garden. One person, a wheelchair user, said:

I never had a chance to look down on people, I am constantly looking up, and what would be great in the garden is to have the chance to be the higher up one. So in the design of the garden there is a raised platform that was partly put in there for that reason, so that as a wheelchair user one can have a higher vantage point across the garden (Royal Institute of British Architects 2009b).

Another example of feedback from consultation with individuals with disabilities was concerned with tackling the sharp drops on steep slopes by using off-cuts of trees and stones as a defence against going over the edge.

When we first got to Eden our first trip was from the top of the hill to the bottom. Two of us came, I was in an electric chair and I have a friend who works with me and was in a manual chair, and the first thing we really got was the fact that as we were getting down the hill, there were very sharp drops down the sides, and then we came to the conclusion that off-cuts of trees and stones could be used as a defence against going over the edge, and there could be little edges as well so even if we go over the edge, we have a soft landing (Royal Institute of British Architects 2009b).

In addition, the Eden Project gardens, acknowledging that the site could be challenging for older people, wheelchair users and young children, provide easy-access buses directly linked to mainline stations, as well as a land train that is accessible for wheelchair users and takes people from the visitors' entrance right down to the bottom of the site and back up again.

The Eden Project gardens also include the Core Building, which is an educational centre located at the heart of the main site. The Sensory Trust, who were involved in the planning and designing and throughout the construction phases of the building, were keen to integrate the inclusive design into the whole planning process. Hence one of the main things they looked at was provision of ground-level emergency egress along the three floor levels instead of using the emergency staircase exit.

We wanted to have a building where anybody using it would feel they are entering and leaving it on an equal basis, and what we achieved through landscape architects and architects working very closely together with the whole of the design team was to look at a creative way of using land form as a way of ensuring that, really, whatever level you are on in the building, you are able to go out at ground level outside. We also have a lift in place, but the other thing we realised is that a lift can breakdown, but if the lift breaks down in this building, it kind of doesn't matter because if you are a wheelchair user you can go around the outside. We realised it made for a more beautiful building, and the integration between the outside and inside, you do not notice it (Royal Institute of British Architects 2009b).

Signage and written information are other features that the Eden Project gardens sought to enhance. Recognising that there are around seven million people in the UK who do not have basic literacy skills and around one million with learning difficulties, the Eden Project gardens used pictures to accompany text, so all users, including those who do not understand English, can comprehend the content. Moreover, signage and waymarkings have been carefully designed with adequate visual contrast and clear text, which enables visitors to find their way around.

Management is another feature that was looked at in the Eden Project gardens. Organising a music session during the summer each year, the Eden Project gardens invited some people to attend specifically to evaluate the event in order to enhance and improve the services.

People with assistance dogs, those with sensory impairments, wheelchair users and people with learning disabilities shared their feedback on what they liked and what sort of access improvements they would like to see. This type of constructive feedback has been used over 3 years to improve the Eden Project gardens so they become more inclusive. These gardens, having adopted the ethos of inclusive design, were voted worthy of inclusion in the New Rough Guide to Accessible Britain in the parks and gardens section in 2010.

An example of inclusive design … we made handrails to improve access, and we made them out of material that would been local to the Mediterranean, sweet chestnut, and they have turned into a wonderful tactile experience, so the handrails double up not just as access aids, but also thinking about other details and making things nice to touch. Another thing is to do with seat heights. I am not particularly tall and standard seat heights are not comfortable so what we tried to create is seats with different heights, and that is a really important thing to think about throughout the design process (Royal Institute of British Architects 2009b).

3.2.5.5   The Roundhouse

The Roundhouse is classified on the UK scale of architectural quality and historical interest as a Grade II listed building, which means that the building is of outstanding importance and of exceptional interest (Bright and Cook 2010, p. 151). This case study is an example of an existing Grade II listed building going through an extension phase.

Located in north London, the Roundhouse is an iconic building, which was built in 1846 as an engine maintenance shed, and is one of the first and the best surviving in the country. In 1854, the Roundhouse was converted into a bonded warehouse for wine and spirits, which was its function until 1960. In the early 1960s, Arnold Wesker took possession of the building, established the Centre 42 Theatre Company and adapted the building into a theatre. In 1983, the Roundhouse was closed and was placed on the English Heritage Buildings at Risk list until it was purchased by Torquil Norman in 1996. Norman aimed to create both a world-class performance venue and a place that would encourage the involvement of the local community. Since then the building has been transformed into an accessible and revitalised structure, which not only retains the beauty and elegance offered by the constructional details of the original space, but also houses a fully converted basement area containing practice rooms, recording studios and social spaces for young people to develop their creative talents (Bright and Cook 2010, pp. 151–152).

> We wanted to transcend physical accessibility by miles. It has to become psychologically accessible both in terms of how you feel about it through its programmes and what it does, but also so that you can see into it, and you do not have to a degree in modern art to get into it. It is a totally democratic feel to the space (Royal Institute of British Architects 2009c).

The executive and board committee of the Roundhouse recruited John McAslan & Partners architects and David Bonnet Associates, an access consultant team, to form an advisory board of people with different abilities to engage in the process of converting the building to an accessible and inclusive space.

Essential to catering for a wider range of people with different abilities was the need for the architects to understand the ambition of the owner of the building, the requirements of the general users and then the necessity for users to test the building.

> What architects are not used to is their propositions being tested by users. Now I think one of the extraordinary things about so-called access groups is that they are people who have particular requirements, they are a testing mechanism at an early stage for the architect to say, "This is what we are planning here, what do you think?", and if you get a constructive dialogue going there, the architect benefits hugely from that exposure because problems may emerge that would simply not have been thought about before, and so everybody should be thanked for actually speeding up the process of design rather than slowing it down (Royal Institute of British Architects 2009c).

To identify the building's limitations and opportunities, access consultants at the Roundhouse carefully analysed the building and proposed solutions to cater for all users, including disabled people. One example of a solution is the colour-coded orientation guidance provided in the circular building, designed to assist users to find their way around and to know which level they are on. Another example is the way that access consultants tackled the issue of having narrow arches in the lower part of the building by specifying that some wider arches should be used in the main approaches so all users, both wheelchair users and ambulant people, could benefit.

> In the lower part of the building there are arches, and there was a deep concern about them being made wider in order to get wheelchairs through or people who walk with sticks. If we look closely at the building plans, some corridors are sufficiently wide and others are not, and so you identify those routes that are wider (Royal Institute of British Architects 2009c).

In addition, architects aiming to achieve complete accessibility started to extrapolate from the historic element of the building all the vertical and horizontal connections that allowed access to the space.

> In the main auditorium space, we have a vintage gallery level, and all the access to this is in the new wing, that allowed for great flexibility in the design and also in terms of reducing the impact of the building itself (Royal Institute of British Architects 2009c).

To achieve inclusion, architects and access consultants focused on catering for the users' needs by providing lower urinals, clear signage and little details, such as handrails on staircases that were carefully designed so that they contrast visually with the wall and background colour, they turn down so they do not catch on people's clothing, and they turn to the wall so that people can see, hold and feel them easily.

> All those kinds of things can make the building accessible. You can make the building physically accessible by having level access, but it is how you interface with the building, because if you have a white building with white walls and white floors, people are not going to find their way around. The signage was a real triumph here, we went through stages where we had the access user groups coming along to look at prototypes, and now our signs that we see in the building reflect the consultation with the user groups (Royal Institute of British Architects 2009c).

The café at the entrance level illustrates good inclusive design in its wide openings that do not restrict the movement of people, its flexibility in providing different types of tables and chairs, and its location close to the entrance, lift and toilets.

> The tables and chairs, instead of being cluttered up near the bar, where you want to get service and place an order, are kept away from there. The location of the café is by the entrance; that is how you capture your café people and enable them to spend their money. It is by the entrance, by the WC and by the lifts, so it makes a tidy package as a designer and makes sense as a user, and it makes the best business opportunity for the café because it has to earn people's goodwill. And also we have a range of tables in that café: there are some tables where you can get your knees under very easily with a central pillar and there are other tables that have a leg at each corner, which are much more stable, and there are some with benches, and there is a range of choices. And I think a lot of what access is about is having choices available to you so you can use your own common sense to say, 'I want to be here rather than there' (Royal Institute of British Architects 2009c).

Consultation played an important role in enhancing accessibility and creating an inclusive design at the Roundhouse. An accessible toilet is an example of how such consultation helped in providing a facility toilet that all users could benefit from.

> For wheelchair users, an accessible toilet is a matter of space and accessibility in getting into and off the toilet; for partially sighted people it is a matter of visual contrast; people with arthritic problems will find screw taps impossible to use. And so actually you have a lot of requirements, and that is why you need a range of people with a range of impairments on your disabled consultation group so they can give you all of that information (Royal Institute of British Architects 2009c).

Although the Roundhouse is located on a hill, with a few steps up to its front entrance, it provides a ramp alongside those steps as an alternative entrance. Moreover, the floors were carefully designed so they are level in order to reconcile the external levels with the internal ones.

Dr. David Bonnett from David Bonnett Associates stated: 'The entrance to a building is the most important part of it, because if I can't find my entrance to the building, then there will be no point of going there'.

One visitor with visual impairment shared his experience of accessing the building:

> In the Roundhouse I can find the texture at the top of the steps, I can find the handrail and then go down the steps and that will guide me eventually to the entrance where I hear the door open, and I know as I go ahead that I walk straight, I come to the box office and there is a person who says: 'Yes, sir, would you like to buy a ticket?' (Royal Institute of British Architects 2009d).

3.3   Inclusive Design as Mainstream

The above-reviewed inclusive case studies have shown that the inclusive design approach is shifting towards the mainstream. It is important to note that what distinguishes inclusive design in the four case studies is that it is invisible and subtle, and ultimately it provides a space that all will benefit from. Although there was a tendency in the past to provide specific services and facilities for individuals with disabilities, legislation, market demand and demographic changes are playing a role in mainstreaming inclusive design.

> There was a tendency, I think, in the past when accessibility was a new thing, people wanted to proclaim it a little bit too loudly and there were unnecessary colours or evidence of certain features, really, in essence, to demonstrate the innovative thinking of the client or the designer. I think we are starting to move beyond that now, where in this building the access features are actually simply part of the mainstream, part of the fabric, they are very subtle, they don't need to be overly proclaimed.

> But, in the main, most features to do with accessibility suit most users actually, so why make them special? So I think the direction things are going in is to get away from the so-called 'special cycle' and simply mainstream it : that is what most disabled people want, quite candidly, and I think, particularly younger people here, they don't want to be singled out, they come here to be cool and they want to be cool (Royal Institute of British Architects 2009d).

This chapter presents four different case studies that have adopted the inclusive design approach, of which the Carrington Building at the University of Reading and the Willows School are two examples of educational buildings. The Willows School case study is an example of an educational building that brings together a special school, primary school and community facilities into single-shared building and campus, whilst the Carrington Building at the University of Reading is an example of how the built environment can tackle accessibility, environmental and sustainability issues. These two educational case studies can be used as references to resolve similar accessibility barriers identified at the University of Kent.

In addition, the University of Arizona case study can be used to investigate whether achieving an inclusive built environment at the University of Kent can overcome the same barriers found on the University of Arizona campus.

The Eden Project gardens case study is an example of how inclusive design can tackle landscape and external environments to achieve an outdoor environment that can anticipate the wide range of users' abilities and needs. Moreover, the Eden Project gardens, as illustrated by the Core Building, show how internal and external environments can be integrated to provide emergency exits that afford a level-access route that is beneficial for all potential users. Hence, the Eden Project gardens is a good case study for use as a reference with respect to the landscaping and external environments at the University of Kent.

Finally, the Roundhouse is a case study that shows how an existing Grade II listed building can achieve inclusive design when undergoing an extension phase. It can be used to propose similar design solutions to tackle accessibility barriers and preserve the heritage and historic character of existing historic buildings in university environments.

The inclusive case studies reviewed in this chapter emphasise the importance of inclusive design in creating built environments that anticipate the needs of all potential users, including individuals with disabilities, thus allowing them to become integrated into society.

This book, in discussing the inclusive design principles and approach, aims to promote inclusive design so that it becomes part of the mainstream. To achieve this, the case studies presented above aimed to identify the architectural and management barriers likely to be encountered in university buildings by using inclusive design methods, which involve consultation with individuals with disabilities, education providers and architects in addition to assessment of the physical environment.

References

Access Committee for England. (1995). *Response from ACE concerning the DOE's proposals for Part M, New dwellings*. London: ACE.

Bright, K., & Cook, J. (2010). *The Colour, light and contrast manual: designing and managing inclusive built environments*. Oxford: Wiley Blackwell.

British Standards Institute. (2005). *New British Standard addresses the need for inclusive design*. Retrieved July 20, 2018, from https://www.bsigroup.com/en-GB/about-bsi/media-centre/press-releases/2005/2/New-British-Standard-addresses-the-need-for-inclusive-design//

Building Control Guidance (2006). *Building control guidance note*. Available at: [Accessed 28 July 2019].

Building Control Guidance. (2016). *Building control guidance note*. Retrieved July 20, 2018, from https://www.charnwood.gov.uk/pages/building_control_guidance_sheets

Burton, E., & Mitchell, L. (2007). *Inclusive urban design: Streets for life*. Oxford: Architectural Press.

CABE (Commission for Architecture and the Built Environment) (2008). *The principles of inclusive design*. London. Retrieved July 20, 2018, from https://www.designcouncil.org.uk/sites/default/files/asset/document/the-principles-of-inclusive-design.pdf

Clarkson, J., Coleman, R., Keates, S., & Lebbon, C. (Eds.). (2003). *Inclusive design: Design for the whole population*. London: Springer Verlag.

CUD (Center for Universal Design). (2008). *The principles of universal design*. North Carolina State University. Retrieved July 20, 2018, from https://projects.ncsu.edu/design/cud/pubs_p/docs/poster.pdf

Department for Communities and Local Government. (2003). *Planning and access for disabled people: a good practice guide*. Retrieved July 30, 2018, from http://www.communities.gov.uk/documents/planningandbuilding/pdf/156681.pdf

Goldsmith, S. (1997). *Designing for the disabled: The new paradigm*. Oxford: Architectural Press.

Goldsmith, S. (1998). Delays to disabled access. *The Architect's Journal, 208*(9), 63–65.

Goldsmith, S. (2001). *Universal design: A manual of practical guidance for architects*. Oxford: Architectural Press.

Imrie, R., & Hall, P. (2001). *Inclusive design: Designing and developing accessible environments*. London: Spon Press.

Inalhan, G. (2012). Teaching inclusive design: Can place-based learning provide access by design? *Access by Design, Autumn, 132*, 29.

Levine, D. (2003). *Universal design New York 2*. New York: Center for Inclusive Design and Environmental Access. University at Buffalo, State University of New York. Retrieved July 30, 2018, from www.nyc.gov/html/ddc/downloads/pdf/udny/udny2.pdf.

Lifchez, R. (1987). *Rethinking architecture: Design students and physically disabled people*. London: University of California Press.

Mace, R. (1985). *Universal design, Barrier free environments for everyone*. Los Angeles: Designers West.

Marrow, R., Bright, K., Kelly, M., Manley, S., Moore, K., Ormerod, M., Walker, A., & Ostroff, E. (2003). *Building & sustaining a learning environment for inclusive design: A framework for teaching inclusive design within built environment course in the UK*. Retrieved July 30, 2018, from http://cebe.cf.ac.uk/learning/sig/inclusive/full_report.pdf

Nussbaumer, L. (2012). *Inclusive design: A universal need*. New York: Fairchild. London: Bloomsbury.

Ormerod, M., Newton, R., Morrow, R., & Thomas, P. (2002). *Inclusive design and the built environment*. In *What is the built environment for if not for people?* Birmingham: City Limited.

Rattray, N., Raskin, S., & Cimino, J. (2008). Participatory research on universal design and accessible space at the University of Arizona. *Disability Studies Quarterly, 28*, 4. Retrieved July 30, 2018, from http://dsq-sds.org/article/view/159/159.

RIBA (Royal Institute of British Architects). (2009a). *Inclusive design (Clip 6, 7 of 7). Inclusive design: Creating a user's world: Part 3: The Willow school*. [Videofile]. Retrieved July 30, 2018, from https://www.youtube.com/watch?v=-z82Ql3cGK4. https://www.youtube.com/watch?v=6Omb6k3EarI

RIBA (Royal Institute of British Architects). (2009b). *Inclusive design (Clip 4, 5 of 7). Inclusive design: Creating a user's world. Part two: Eden Project: Working with the Sensory Trust*. [Videofile]. Retrieved July 30, 2018, from http://www.youtube.com/watch?v=fTRq1oIlT0Y&feature=relmfu. https://www.youtube.com/watch?v=DfnLdXYVLc8

RIBA (Royal Institute of British Architects). (2009c). *Inclusive design (Clip 2 of 7). Inclusive design: Creating a user's world. Part one: The Round house*. [Videofile]. https://www.youtube.com/watch?v=1K2bRJFvkJY

RIBA (Royal Institute of British Architects). (2009d). *Inclusive design (Clip 3 of 7). Inclusive design: Creating a user's world. Part one: The Round house*. [Videofile]. Retrieved July 30, 2018, from https://www.youtube.com/watch?v=IZSJafFwIr0&t=24s

Sawyer, A., & Bright, K. (2007). *The access manual: Auditing and managing inclusive built environments*. Oxford: Blackwell Publishing Inc..

Sommer, R. (1983). *Social design: Creating buildings with people in mind*. Englewood Cliffs, NJ: Prentice Hall.

Steinfeld, E. (2010). *Inclusive housing: A pattern book, design for diversity and equality*. New York: W.W. Norton & Company, Inc..

Steinfeld, E., & Danford, G. S. (1998). *Measuring enabling environments: Measuring the impact of environment on disability and rehabilitation*. New York: Llumer Academic Plenum Publishers.

Steinfeld, E., & Maisel, J. (2012). *Universal design: Creating inclusive environments*. Hoboken, NJ: John Wiley & Sons.

Story, M. F. (2001). *Principles of Universal Design*. McGraw-Hill. Takahashi: *Universal Design Handbook*.

Waller, S., & Clarkson, P. J. (2009). Tools for inclusive design. In C. Stephandis (Ed.), *The universal access handbook*. Boca Raton, FL: Taylor & Francis.

Chapter 4
Researching Inclusive Design at Universities: The University of Kent Case Study

4.1 Introduction

The study, which is the subject of this book, was a qualitative and quantitative one reviewing critical architectural barriers that limit accessibility for potential users at the University of Kent Canterbury. It took the inclusive design approach as a model in order to enhance accessibility and promote inclusion for all potential users, including individuals with disabilities.

This chapter justifies the case study method, based on a multiple-case-study design, as the one most applicable to the study. One of the purposes of the study was to sharpen our insights into, and awareness of, the architectural needs of a broad spectrum of users, including individuals with disabilities, in order to develop design strategies and recommendations to satisfy the needs of all users.

A triangulated approach to data collection was utilised. More specifically, a number of different methods and assessments were carried out to identify the architectural barriers and users' level of satisfaction with the services provided at the University of Kent. The study also included an access audit assessment, in which individuals with disabilities helped to assess the physical environment and highlight the barriers they encountered; architectural building plans were analysed; real-time observation was carried out; and semi-structured interviews were conducted with students with disabilities and staff members, education providers and architects at the university.

The study explored the pedagogical perspectives and implicit theories used to identify the architectural and management services provided at this university to cater for the needs of potential users, both those with and those without disabilities. The study focused on a series of specific questions to investigate whether adopting an inclusive design approach in a university setting is preferable to just meeting legal building requirements. The questions were as follows:

1. What are the physical and management barriers which hinder individuals with disabilities from accessing the university?
2. To what extent does the University of Kent, Canterbury, comply with the Building Regulations to influence access and eliminate architectural barriers?

© Springer Nature Switzerland AG 2020

I. Shuayb, *Inclusive University Built Environments*,
https://doi.org/10.1007/978-3-030-35861-7_4

3. To what extent are students, with or without disabilities, and staff members satisfied with the services provided?
4. Which inclusive/accessible approach would users most like to see adopted at the University of Kent, Canterbury?
5. To what extent have education providers managed to make reasonable adjustments or eliminate architectural barriers?
6. To what extent are architects aware of the needs of potential users and people with disabilities when designing university buildings?
7. To what extent are architects aware of the inclusive design approach?
8. To what extent do architects abide by the design rules and regulations in their designs to cater for all users, including people with disabilities?
9. Has the guidance influenced architects' designs of new buildings or limited their creativity?

## 4.2	The Study Framework

### 4.2.1	Mixed-Methods Approach

Although a multiple-case-study approach is well established in the qualitative research tradition, this study adopted a mixed-methods approach that explored the aims of the research through a combination of quantitative and qualitative measures. Gall et al. (2010) define a mixed-methods research study as 'a type of study that uses both quantitative and qualitative techniques for data collection and analysis, either simultaneously or consecutively, to address the same or related research questions' (p. 461). Mixed-methods designs include methods from both the qualitative and quantitative research traditions but join them in a distinctive technique to answer research questions that could not be answered if only one way were used (Tashakkori and Teddlie 2003). Hence the choice of research methods depends upon the questions to be asked, and the questions depend on their context (Denzin and Lincoln 2011).

One of the aims was to explore the experiences of individuals with disabilities in accessing the built environment at the University of Kent and their level of satisfaction with the services provided. The study also used qualitative measures to survey the current practices used to enhance the level of accessibility to the university buildings. Finally, qualitative and quantitative methods were used to investigate what service options for end users are offered and what are the physical and management barriers that people with and without disabilities encounter when accessing the built environment at the University of Kent.

The inclusion of qualitative analyses in this study was regarded as 'essential to make sense of actual lived experience' (Marecek et al. 1997, p. 632). In qualitative research, the researcher is the primary instrument for data collection and analysis (Al-Hroub 2015; Winegardner 2001). The researcher assesses interview participants, obtains answers relevant to his/her research questions and observes users in their natural setting. Moreover, a qualitative study contains descriptive data and information that is articulated in words rather than numbers. A participant may include information about personal experiences, problems and concerns, and preferences described in his/her own words which can be used as evidence-based data by the researcher. Some qualitative research can be distinctive from the generally known forms of qualitative studies as it is designed to respond to the study's changing conditions as it develops. This type of qualitative study usually occurs with a small, non-random and purposive sample, which usually takes an extensive amount of time and enables the researcher to be in close contact with the research participants (Al-Hroub 2014; Winegardner 2001). Maxwell (2016) lists five research purposes to which qualitative studies are particularly suited:

1. Understanding the meaning for the participants in a study of the events, situations and actions with which they are involved, and the accounts they give of their lives and experiences.
2. Understanding the particular context within which the participants act, and the influence this context has on their actions.
3. Identifying unanticipated phenomena and influences, and generating new, grounded theories about them.
4. Understanding the processes through which events and actions take place.
5. Developing causal explanations: This matches the significance of studying the specific needs of all users in order to understand the different views of users, education providers and architects when it comes to tackling these needs and taking account of their views within the field of inclusive design (Maxwell 2016, p. 221).

Quantitative methods were appropriate for measuring both students' and staff members' attitudes towards, and levels of satisfaction with, the built environment in terms of catering for their needs. Therefore, it was also a valuable research method for determining the architectural barriers, and identifying the preferences or beliefs of this population with regard to the inclusive design approach. According to McCullough (1997), quantitative methods have some advantages. One of the advantages is that the results of the study are statistically reliable since the obtained data confirms that a concept, theme or idea is favoured to other alternative ones. Moreover, the obtained data can be generalized to the participants whose responses are similar to the total participants who have been surveyed (Mcullough 1997). Edwards (1998) adds that quantitative multivariate methods allow researchers to measure and control variables.

Summing up, both qualitative and quantitative research methods were able to make valuable contributions to the present study. Each used method had a different contribution that enabled the researcher to find answers for the conducted research study. They should be considered as complementary, not opposing, techniques of analysing the data, and which method or technique is chosen depends on which one enables the researcher to find an answer for a specific research question, as well as which is more comprehensive, clearer, more complete and, above all, more descriptive of reality.

4.2.2 *Validity and Reliability of Case Study*

Whilst there are no specific criteria to measure validity and reliability for case studies, validity commonly represents the accuracy and value of the interpreted data and findings, whilst reliability measures the degree to which other researchers would conclude the same findings if they were using exactly the same research tools and instruments. In the present study validity was determined by the following means:

- Researcher positioning: The researcher showed sensitivity by relating to the issue and topic being researched.
- Triangulation: use of multi- and mixed data collection methods, different sources, access audit assessments, observation, interviews and artefacts in conjunction with analysis in order to check validity.
- Member checking: confirmation of data by students with disabilities and able-bodied students and staff members, education providers and architects.
- Pattern: noting the frequency of patterns identified in case study data to check its association and correlation with previous research findings.
- Long-term assessment and observation: data collection over a period of 12 months and regular consultation with the participants.
- Coding check: quantitative methods are used to check the reliability of coded data during the data analysis phase.

Validity and reliability were both examined in the present study to validate the accuracy of data findings. 'Reliability is less a function of replicability and more one of the credibility of the researcher's claims to knowledge and acknowledgment of his/her central role, relationship, and biases in the research' (Merriam 1998). On the other hand, the data collected using the qualitative method is considered rigorous by its nature as it comes from 'the researcher's presence, the nature of the interaction between researcher and participants, the triangulation of data, the interpretation of perceptions, and rich, thick description' (Merriam 1998). Yin (2009) suggests researchers to utilize several and different evidence-based measurements to ensure that the collected data is validated, and this was done in the present research. Yin (1994) asserts, 'external validity could be derived from theoretical relationships, and that from these, generalisations could be made. It is the development of a formal case study protocol that provides the reliability that is required of all research'.

4.3 Study Design

The primary intent of this study was to investigate whether adopting an inclusive approach at the University of Kent was preferable to just meeting legal building requirements. The reason for choosing a university as a case study rather than a school was that higher educational institutions play a major role in providing the professional training for high-level jobs, as well as the education necessary for the development of the personality of all their students including individuals with disabilities. Moreover, universities have an important part to play in promoting social inclusion and participation in the mainstream society.

Acknowledging that antidiscrimination disability legislation ensures and promotes the full realisation of all human rights and fundamental freedoms for individuals with disabilities to enable them to attain equal access to higher education and employment, this study aimed to investigate whether the University of Kent has indeed managed to eliminate barriers in order to ensure that individuals with disabilities have equal access to higher education which will then enable them later to gain access to employment.

Developing an inclusive built environment for all potential users is a key aspect of the whole university policy not only in the UK, but also in other European countries. Inclusivity embraces not only people with disabilities, but also all other users and age groups. A key finding from the review of previous case studies revealed that although many academic and scientific projects and studies have been carried out to enhance accessibility of products, few studies have been conducted on the built environment to investigate the level of accessibility and whether inclusive design principles have been applied.

The study carried out physical assessments of six selected buildings at the University of Kent to investigate the level of accessibility and the ways in which these buildings accommodate the needs of individuals with disabilities. Consultation with students and staff members with disabilities in which they shared their experiences of accessing the built environment was important when it came to highlighting accessibility issues.

Moreover, personal interviews were conducted with commissioned architects at the University of Kent to investigate their knowledge and understanding of the needs of disabled people during the design and implementation phases. Similarly, personal interviews were carried out with education providers at the University to determine their input in complying with their legal duties to remove physical barriers and promote equality and diversity.

The University of Kent was selected as the case study because:

1. It has one main campus at Canterbury.
2. It owns its own buildings, which facilitated the carrying out of the access audits and observation of the buildings in use.
3. The University of Kent gave permission for the study to be conducted within a feasible time frame.
4. Online surveys were a vital means of data collection, and it was important that the case study university could provide suitable online tools in order to reach a large population of students and staff members easily. The University of Kent provided and facilitated contact by allowing their websites to be used to circulate surveys among the university population. This was handled by the Student Records Office at the University.

5. The researcher was able to reside in close geographical proximity to the above selected university. This made it possible to conduct the research most effectively in terms of time and logistics. All participant interviews were held in an accessible and friendly environment at a university venue.
6. The selected university contained a good number of students, which made it possible to choose the sample from as large a population as possible.

4.4   Research Study Sample

A group of individuals with disabilities, namely mobility, visual and mental impairments, was selected from the University of Kent. Face-to-face interviews were carried out with ten participants in order to capture their experiences of accessing the University's buildings and their levels of satisfaction with the services provided.

The criteria used to select the students at the University of Kent were based on the purposive sampling selection technique, which is used when the population is too small to obtain a random sample (Tongco 2007). 'Purposive sampling selection is a type of non-probability sampling that is most effective when one needs to study a certain cultural domain that includes knowledgeable experts' (Tongco 2007). The study started with an online survey, and then a purposive sample was selected based on the survey (Tongco 2007). Participants who showed interest in taking part in a personal interview were contacted via email to arrange a meeting.

4.5   Data Collection Methods

Two data collection phases were used to address critical issues associated with the process of identifying and recommending an inclusive design proposal to tackle the architectural barriers that hindered individuals with disabilities from accessing the built environment at the University of Kent. The data collection phases involved conducting an online survey, carrying out access audits on six buildings, and personal interviews with ten participants with disabilities, four education providers and two architects at the University of Kent.

The phases also encompassed feeding back the results and consulting participants with disabilities. The first data phase explored how the level of accessibility at six selected buildings was identified and assessed. The second data phase involved analysing the findings of the first phase and the artefacts/architectural plans of four buildings which findings from the first phase revealed as being the most problematic in terms of promoting segregation and exclusion.

4.5.1   First Data Collection Phase

4.5.1.1   Online Survey

An online survey was the first method to be used. The purpose of the online questionnaire was to determine the views of all users, including those with disabilities, and to obtain details of their experiences of accessing the built environment at their university. The online survey contained 24 questions, divided into three sections and consisting of multiple-choice and free-text answers. The first section requested personal background information and contained questions about types of disability and their occurrence, and levels of satisfaction with British disability legislation. The second section aimed to collect information about the level of accessibility to buildings and barriers encountered, and included questions about participants' experience of accessing the built environment, means of transportation used, external and internal features of buildings, signage and emergency exit routes (Appendix B). The last section was directed at those willing to take part in personal interviews and consultations, and hence it asked for personal contact details.

The data for the online survey at the University of Kent were collected over three academic terms and circulated among a representative sample of 1000 non-disabled people and 1984 individuals with disabilities registered at the Disability and Dyslexia Student Services Department.

4.5.1.2   Access Audits

Access audits were the second method used to collect data. An access audit is 'the process of measuring the accessibility and usability of services and facilities against predetermined criteria. Its aim is to identify physical barriers and consider means of eliminating or mitigating them' (CAE 2005). The purpose of an access audit is thus to provide information about the accessibility level to a building and allow the auditor to allocate both the accessible and inaccessible facilities to enable him/her later to propose design solutions and improve access for potential users (Sawyer and Bright 2007). In the present study, the audit investigated the compliances of selected buildings at the University of Kent with the British Building Regulations 2000 Part M, Access to and use of buildings, and Approved Document M 2004 to determine the level of accessibility for potential users, including those with disabilities.

Access audits were carried out on six selected buildings at the University of Kent. The selection was driven by the need to evaluate the main buildings and facilities used by students, in addition to appraising recently built or modified buildings. A shortlist was developed following consultation with students who completed the online survey at the University of Kent, leading to four buildings being selected. These four were the Templeman Library, Eliot College, the Registry and the Venue, the student union nightclub. In addition, the Marlowe Building, built in 1960, was selected as an example of an older building dating from the 1960s when the

University was founded, whilst the new Jarman Building, the School of Arts, was chosen as an example of a new constructed building, making six in total.

The access audit looked at two main features, physical features and management procedures.

The following equipment were used to assess accessibility to the selected buildings:

1. Measuring tape: used to determine the current size of facilities. The data obtained were used to make comparison between the required and actual measurements to determine whether or not they complied with Part M standards.
2. Gradient level: used to determine the gradient of external and internal ramps or any sloping surface. The data obtained were used to conduct a comparison between the required and actual measurements to determine compliance with Part M standards.
3. Manual spring balance: a Super Samson Salter Spring Balance was used to determine the door opening force, which again was used to compare the required and actual force to determine compliance with Part M standards.
4. Digital camera: used to document accessibility conditions during the access audit, as well as to record the management procedures and practices used. Photographs were included in the audit reports and when analysing the findings.
5. Sketchbook: used to record data manually by drawing conceptual diagrams of movement of people and the building layout, as well as recording measurements and problems that were identified via observation.

To investigate the barriers encountered at each selected building, a journey was considered to be a complete cycle, starting with the external environment of a designated building, moving on to the main entrance and then inside the building, the use of its facilities and also evacuation of the building in case of emergency.

This assessment permitted an appraisal of the route that a wide range of potential users might follow to reach a desired destination. The audits were designed to take into account the journey from the user's means of transport to reach the university, and on to the final destination, taking into account factors such as signposts, design of pedestrian environments, bus stops, and car parking spaces and major changes in level (Fig. 4.1).

The author, in the role of auditor, conducted a thorough assessment of the accessibility of buildings. The assessment included taking notes, critical measurements of design features where appropriate and photographs of features of particular relevance. She also gathered real-time information about operational aspects and staff attitudes whilst observing buildings in service.

Adopting the inclusive design principles, the auditor studied how well each feature satisfied the accessibility requirements of the following users:

- Those with mobility difficulties/wheelchair users
- Those with visual impairments
- Those with hearing impairments
- Those with multiple disability

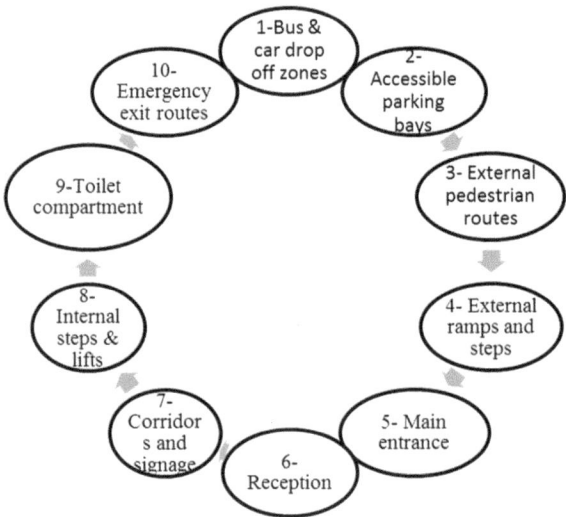

Fig. 4.1 The access audit journey cycle

- Those with a specific learning disability/(dyslexia)
- Those with mental health difficulty
- Those with hidden disability (e.g. diabetes, epilepsy, heart condition)
- Pregnant women
- Mothers/carers with a pushchair

On completion of the audit, an access report and a summary were produced for each building. The audit report presented in a tabular document provided detailed descriptions of features of each building and suggestions for improving access. Each feature was rated in accordance with the inclusive design criteria used in Table 4.1.

4.5.1.3   Interviews with Students and Staff Members with Disabilities

Individual face-to-face interviews were conducted in this study, and these were semi-structured (Groat and Wang 2002). The purpose was to obtain details of participants' experiences and views, and an indication of their satisfaction or otherwise with the services provided in the built environment. An email was sent to the ten volunteer participants with a research brief and a consent form containing research background information about the research, selection criteria, details about the person conducting the interviews, venue, research risks and benefits, and confidentiality and analysis procedures used.

The participants were asked to comment on their recent experience of accessing the built environment at the University of Kent, Canterbury campus, with sessions structured around key stages in the journey cycle (Appendix C). All interviews were audio-taped and their duration ranged from 45 to 60 min. In addition, four participants

Table 4.1 Rating system used for audit summary document

Excellent	The criterion refers to the needs and wants of potential users. Offers diversity and difference; Relates to the requirements of accessibility standards and guidelines in Approved Document M; Relates to health, safety and welfare.	****
Very good	The criterion refers to the needs and wants of potential users; Relates to the requirements of accessibility standards and guidelines in Approved Document M.	***
Unsatisfactory	This criterion refers to aspects which: Did not meet the needs and wants of potential users; Did not consider diversity and difference; Did not meet the requirements of several key principles in the accessibility guidelines.	**
Very poor	This criterion refers to aspects which are very distinct from the majority of accessibility guidelines, or which fail to comply with the principles, standards and guidelines.	*

at the University of Kent took part in a site visit to a recently opened building. They included wheelchair users and people with mobility, sensory and cognitive impairments. A comprehensive set of tasks was carried out to ensure that the physical environment and overall journey quality could be assessed.

The author continued the process of interviewing and probing further in accordance with the purpose of the study until she reached a point when the data being collected did not add anything new. The University of Kent interviewees included five postgraduates, three undergraduates, one of whom was both a student and a staff member, and one staff member, all of whom had disabilities which ranged from mobility difficulties, special learning needs, visual impairments, mental health difficulties, dyslexia and multiple disabilities.

4.5.1.4 Interviews with Stakeholders

Semi-structured interviews with stakeholders were conducted as the last part of the first data collection phase. Education providers and architects took part in these interviews, the aim of which was to obtain feedback from a design, operational and commercial perspective regarding the effectiveness of the legislation/design standards and regulations with respect to providing accessible environments and promoting inclusion at the University of Kent. Four education providers were selected, representing departments dealing with student services, architecture project management and emergency escape management. Consequently, the Disability and Dyslexia Student Services (DDSS), Equality and Diversity Department, Estates Department, and Health and Safety Department at the University were all contacted.

The education providers were asked to share their knowledge about meeting the needs of disabled users and addressing the challenges associated with providing an accessible and inclusive built environment. All interviews were audio-taped and the duration ranged from 45 to 60 min.

Furthermore, architects from two architecture firms at the University of Kent were interviewed. The purpose was to investigate their knowledge regarding catering for the needs of disabled users when designing new buildings. The architects were contacted via a similar email to that sent to other interviewees. They were given the chance to review the project brief, ask any questions and indicate whether they would choose to volunteer to take part in the study. All participants were asked to sign consent forms before proceeding with the interviews. All interviews were audio-taped and the duration ranged from 45 to 60 min. Architects were asked to share their knowledge of how to address the needs of all users, the building design standards used and the challenges associated with designing a space catering for all users, including those with disabilities.

4.5.2   Second Data Collection Phase

The second data collection phase included analysis of the findings of the first data collection phase and the artefact/architectural plans of two buildings, which as findings from the first phase indicated were the most problematic in terms of segregation and exclusion.

4.5.2.1   Consultations with Individuals with Disabilities

The second data collection phase involved communicating the findings from the first phase to the previously interviewed participants with disabilities. The author contacted all ten interviewed participants at the University of Kent via email to give them the results and arrange for another meeting. Four participants showed interest in taking part in the consultation session, which involved a presentation showing photographs taken during the access audit of the selected six buildings and questions about the participants' experience of accessing them. Only one participant was able to attend in person to share her experiences, whilst the remaining three did likewise and answered the requested questions in a PowerPoint presentation format via email.

4.5.2.2   Artefacts

It is believed that the architectural plans that related to the audited buildings added to the richness of the data collection. The author was able to obtain the plans of the six selected buildings. She analysed each building plan and investigated the accessibility barriers by identifying the services provided at each building using a colour coding system. The inclusion of artefacts, such as colour coding plans and the master site plan of the Canterbury campus, enhanced the author's understanding of the campus approaches, circulation and functions provided at each building and assisted her in proposing design solutions.

4.6   The Study's Ethical Considerations

The subjects were all treated in accordance with the research ethics requirements. The following ethical issues were addressed in the study:

- The author fully revealed her identity and background to the students and staff members, education providers and architects.
- The purpose and procedures of the study were fully explained to the students, staff members, education providers and architects.
- The author obtained written permission from all the volunteer participants, including students, staff members, education providers and architects.
- Permission was obtained from the Estates Department at the University of Kent to carry out access audits, including taking photographs of the selected buildings.
- The author explained the educational and architectural benefits of the study for the students, staff members, education providers and architects. She also ensured that participants would not be harmed in any way.
- Participants were fully informed that they had the option to refuse to take part and the right to terminate their involvement at any time. This happened when six participants in the Kent case study did not want to participate in the second phase of the research.
- All the information about the participants was treated with the strictest confidentiality. All the participants remained anonymous.
- Participants with disabilities were given a brief results summary after analysing the first phase of data collection for the study.
- All participants were also informed that all audio-recorded interviews would be destroyed following the termination of the analysis.

4.7   Data Analysis Procedure

The data analysis procedure adopted in the case study of the University of Kent analysed 'certain statements and generated themes to reduce the mass of experiences participants had described in order to form a meaningful picture without presuppositions' (Creswell 2017). The author developed a general coding protocol by categorising the emerging data. Literature was consulted, and the author also used ideas that emerged from the individual interviews. She analysed the first-phase data of the case study as soon as they were collected; next she addressed the second-phase data.

Moreover, when the author analysed the data from the first phase, she identified constructs, themes and patterns that best explained the data collected with respect to individual participants and across the whole group. What emerged was coded across segments, with the goal of discovering commonalities that reflected the underlying meaning of, and the relationships between, the coded data, thus making it possible to answer the research questions and fulfil the purpose of the study (Creswell 2017; Corbin and Strauss 2008; Gall et al. 2010).

4.7.1   Individuals with Disabilities, Interviews and Online Survey Analysis

The author developed a general coding protocol by categorising the online data using the SPSS software according to themes and patterns that matched the same themes identified in the interviews with individuals with disabilities. All individual interviews were transcribed, sorted and categorised to identify the constructs, themes and patterns revealed by the participants relating to the architectural barriers, management factors and disability awareness that affected their accessibility.

Each interview was coded using the line-by-line technique. Secondly, the interviews were re-examined using a focused coding approach with the purpose of developing categories. The common patterns were categorised according to six themes that emerged from the data: (1) physical environment barriers, (2) management procedures and practices, (3) staff training and disability awareness, (4) effectiveness of legislation, (5) integration and consultation with disabled people and (6) inclusive approach: design, policies, practices and procedures.

4.7.2   Access Audit Analysis

After completing the access audit of the University of Kent, the author produced access reports and summaries for each university building. Each access audit report was presented in a tabular document providing detailed descriptions of the features of each building together with suggestions for improving access. The second stage of analysis of the access audit reports involved analysing the photographs and highlighting the barriers and then developing common categories and subcategories which were categorised with respect to two themes: (1) physical environment, which contained results relating to external and internal environment features, and (2) management procedures and practices. Table 4.2 illustrates the themes of the categories and subcategories.

4.7.3   Analysis of Stakeholder Interviews

With regard to the results from the stakeholder interviews, education providers and architects helped by providing feedback with respect to identifying the stakeholders' duties in providing accessible environments and eliminating architectural barriers. All individual interviews were transcribed and were sorted and categorised in order to produce themes and patterns which were categorised as follows:

- Compliance with legislation duties
- Management procedures and practices

Table 4.2 Access audit categories' and subcategories' themes

Physical environment	Management procedures and practices

1. External environment
    (a) Reaching the university using the university bus
    (b) Car parking
    (c) External steps
    (d) External ramps
2. Internal environment
    (a) Main entrance
    (b) Reception counter
    (c) Corridors
    (d) Passenger lifts
    (e) Internal staircases
    (f) Toilet compartments
        • Female and male compartments
        • Wheelchair-accessible toilets
    (g) Emergency exit route
	1. Managing the external and internal environment

- Training, and disability awareness
- Integration and consultation with disabled people
- Inclusive approach policies, practices and procedures

4.7.4   Analysis of Artefacts

The second data analysis process involved analysing the architectural plans of each building in order to identify the accessibility barriers by colour coding the services provided in each building. Each facility provided in a given building was given a colour code which helped the author to identify the services and facilities provided there and the circulation routes.

References

Al-Hroub, A. (2014). Perspectives of school dropouts' dilemma in Palestinian refugee camps in Lebanon: An ethnographic study. *International Journal of Educational Development, 35*, 53.

Al-Hroub, A. (2015). Tracking dropout students in Palestinian refugee camps in Lebanon. *Educational Research Quarterly, 38*, 52–79.

CAE (Centre for Accessible Environments). (2005). *Access audit handbook*. London: RIBA Publishing.

Corbin, J., & Strauss, A. (2008). *Basics of qualitative research techniques and procedures for developing grounded theory*. London: Sage Publications.

Creswell, J. (2017). *Qualitative inquiry and research design: Choosing among five traditions* (4th Editors ed.). Thousand Oaks, CA: Sage.

Denzin, N., & Lincoln, Y. (2011). *Handbook of qualitative research in education* (4th Editor ed.). Thousand Oaks, CA: Sage Publications.

Edwards, D. J. (1998). Types of case study work: A conceptual framework for case-based research. *Journal of Humanistic Psychology, 38*(3), 36–71.

Gall, J. P., Gall, M. D., & Borg, W. R. (2010). *Applying educational research: How to read, do, and use research to solve problems in practice* (6th Editor ed.). New York: Allyn & Bacon.

Groat, L.N., & Wang, D. (2002). *Architectural research methods*. Canada: John Wiley & Sons.

Marecek, J., Fine, M., & Kidder, L. H. (1997). Working between worlds: Qualitative methods and social psychology. *Journal of Social Issues, 53*(4), 631–644.

Maxwell, J. A. (2016). *Qualitative research design: An interactive approach* (3rd Editor ed.). Thousand Oaks, CA: Sage.

McCullough, D. (1997). *Quantitative vs. qualitative marketing research*. Retrieved July 30, 2018, from http://www.macroinc.com/english/?s=Quantitative+vs.+qualitative+marketing+research

Merriam, S. B. (1998). *Qualitative research and case study applications in education*. San Francisco: Jossey-Bass Publishers.

Sawyer, A., & Bright, K. (2007). *The access manual: Auditing and managing inclusive built environments*. Oxford: Blackwell Publishing Inc.

Tashakkori, A., & Teddlie, C. (2003). *Handbook of mixed methods in social & behavioral research*. Thousand Oaks, CA: Sage.

Tongco, C. (2007). Purposive sampling as a tool for informant selection. Retrieved July 30, 2018, from http://scholarspace.manoa.hawaii.edu/bitstream/handle/10125/227/I1547-3465-05-147.pdf

Winegardner, K. E. (2001). *The case study method of scholarly research*. The graduate School of America.

Yin, R. (1994). *Case study research: Design and methods* (2nd Editor ed.). Beverly Hills, CA: Sage Publishing.

Yin, R. (2009). *Case study research: Design and methods* (4th Editor ed.). London: Sage Publications.

Chapter 5
Identifying Accessibility Barriers at University Built Environment: Findings from the University of Kent

5.1 Introduction

The purpose of this chapter is to establish the level of accessibility at the University of Kent, Canterbury campus, in order to investigate whether the built environment at the University has adopted inclusive design principles so as to accommodate the needs of all potential users, including people with disabilities.

This chapter reports the access audit findings, which revealed that the University of Kent built environment has many physical and management barriers that limit accessibility for a wide variety of users. Moreover, the chapter reports some of the experiences of students and staff members with disabilities in accessing the built environment and their satisfaction/dissatisfaction with the design and services provided.

The chapter also examines architects' knowledge and their understanding of all users' needs, including those of individuals with disabilities, during the design and implementation phases. In addition, the chapter reports the education providers' input when considering the removal of physical barriers. This chapter investigates the following key questions in relation to the University of Kent:

1. What are the physical and management barriers in the buildings which hinder individuals with disabilities when accessing them?
2. To what extent does the University of Kent comply with the Building Regulations to ensure access and eliminate architectural barriers?
3. To what extent are students, with and without disabilities, and staff members satisfied with the services provided?
4. Which inclusive/accessible approach would users most like to see adopted at the University of Kent, Canterbury?
5. To what extent have education providers managed to make reasonable adjustments or to eliminate architectural barriers?
6. To what extent are architects aware of the needs of potential users and people with disabilities when designing university buildings?

© Springer Nature Switzerland AG 2020
I. Shuayb, *Inclusive University Built Environments*,
https://doi.org/10.1007/978-3-030-35861-7_5

7. To what extent are architects aware of the inclusive design approach?
8. To what extent do architects abide by the design rules and regulations in their designs to cater for all users?
9. Has the guidance influenced architects' designs of new buildings or limited their creativity?

5.2   Scope of the Work

The scope of the study extended to conducting access audits of six buildings. The selection was driven by the need to evaluate the major buildings and facilities used by students, in addition to appraising recently built or modified buildings. A short-list was drawn up, and following consultation with students who completed the online survey, the Templeman Library; Eliot College; the Venue, the student union nightclub; the Registry; the Marlowe Building; and the Jarman Building were selected. As noted above, the author also interviewed a representative sample of students and staff members with disabilities. In addition, education providers and architects were interviewed.

5.3   Approach Taken

The study adopted Finkelstein's (2002) and Swain and French's (2000) social model of disability, which rejects the cultural attitudes and presumptions that consider individuals with disabilities to be victims of having personal tragedy. Instead it puts forward the view that it is the way society is run and organised that is the problem, not the person with the disability. In order to promote social inclusion that takes account of human social lifestyles and healthcare issues, architectural and attitudinal barriers have to be removed.

Consequently, to investigate the barriers encountered in each building, a journey was considered to be one complete cycle, starting with the external environment of a designated building, moving on to the main entrance, and then inside the building and the use of its facilities inside, and also evacuation of the building in case of emergency. This assessment permitted an appraisal of the route that a wide range of potential users follow to reach their desired destination (Fig. 5.1).

Fig. 5.1 The journey cycle

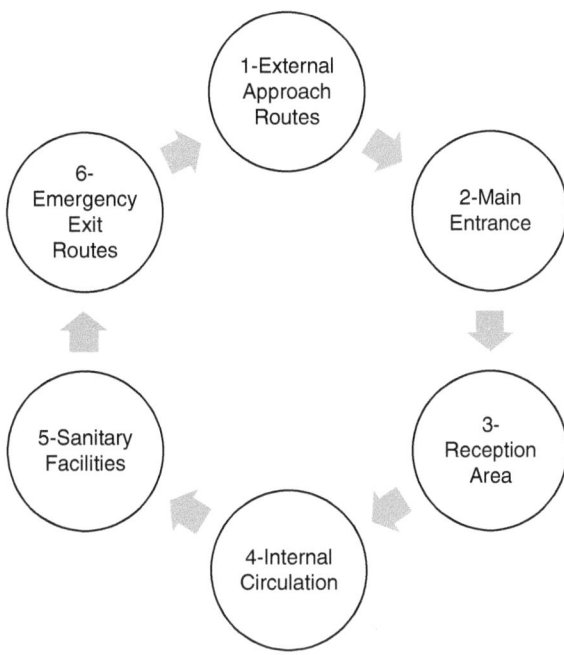

5.4 Methodology

5.4.1 *Access Audits*

The access audits were based on guidance in Approved Document M of the Part M Building Regulation, 2004. The auditor assessed the building and services with respect to a number of regulatory and best practice guidelines. The access audit looked at two main features: physical and management.

5.4.2 *Students and Staff Members with Disabilities*

The purpose of this part of the project was to obtain data on the experiences, views and satisfaction with the level of accessibility to, and the services provided at, the built environment. Ten participants of both sexes and a range of ages and disabilities took part in semi-structured interviews. The participants were asked to comment on their recent experience of accessing the built environment at the Canterbury campus, with sessions structured around key stages in the journey cycle. In addition, four participants, one wheelchair user, one with mobility impairment, one with visual impairment and one with mental health difficulties, took part in a site visit to a new building, where a comprehensive set of tasks was undertaken.

5.4.3   Stakeholder Consultations

Education providers and architects took part in this exercise, which aimed to obtain feedback from design, operational and commercial perspectives, with regard to the effectiveness of the legislation/design standards and regulations as they relate to provision of accessible environments and promotion of inclusion.

A total of six semi-structured interviews were held with stakeholders, four education providers and two architects, to consider the effectiveness of the regulations and the challenges associated with providing an accessible and inclusive built environment and educational service.

5.4.4   Online Questionnaire

The purpose of the questionnaire was to ascertain the views of students and staff members and discover their experience of accessing the built environment. The data for this survey were collected over three academic terms. The survey contained 24 questions consisting of multiple-choice and free-text answers. A total of 236 respondents of both sexes and a range of ages, ethnicities and disabilities took part.

5.5   Results

The access audits and consultations suggested that many positive steps had been taken across the university to enhance accessibility. However, the built environment still did not entirely follow inclusive design principles, in that the level of accessibility differed greatly within and between buildings, and in many instances different disability groups, such as wheelchair users, were served more effectively than individuals with visual, hearing and cognitive impairments. These findings suggest that the University of Kent needs to make further changes to make its built environment and building designs more inclusive and user friendly.

The chapter focuses on the analysis of data obtained using the following tools:

1. Access audits
2. Consultations with students and staff members with disabilities
3. Stakeholder consultations

5.5.1   Access Audit Results

5.5.1.1   Overview

Four buildings were selected for audit in accordance with students' proposals: the Templeman Library, Eliot College, the Registry and the Venue, the student union nightclub. In addition the Marlowe Building was selected as an example of an older building and the new Jarman Building School of Arts as an example of a new one. Figure 5.2 is a site plan that shows the access-audited buildings, bus routes and parking spaces.

The access audit investigated the compliances of these buildings with England and Wales Building Regulations 2000 Part M 'Access to and use of buildings', and Approved Document M, 2004. The audits were designed to take into account the journey from the user's means of transport to reach the university, continuing to the final destination, taking into account factors such as signposts, design of pedestrian environments, bus stops, car parking spaces and major level changes. The access audit evaluated two physical environment components, the external environment and the internal environment (Sawyer and Bright 2007). Table 5.1 lists the features that were audited.

The audit involved taking notes, critical measurements of design features where appropriate and photographs of features of particularly relevant aspects in order to investigate whether the University of Kent has adopted the Approved Document M Building Regulation standards to enhance accessibility to its built environment and to promote inclusive environments. The auditor also gathered real-time information about operational aspects and staff attitudes whilst observing buildings in use.

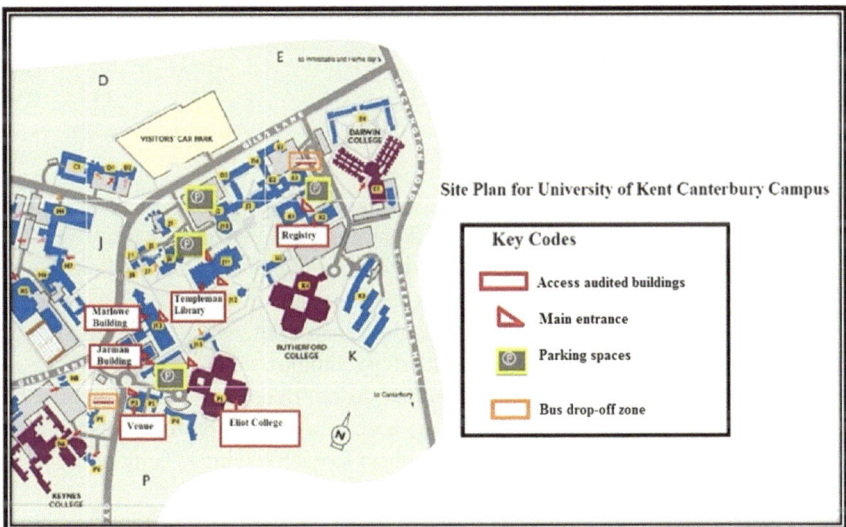

Fig. 5.2 University of Kent site plan

Table 5.1 Features audited

Feature to be audited	
External environment	• Bus drop-off zone • Car parking-accessible parking bays • External steps and ramps
Internal environment	• Main entrance • Reception counter • Doors • Corridors • Elevators/lifts • Steps/stairs • Ramps • Handrails • Toilets • Emergency egress routes

Table 5.2 Number of students by disability classification with percentage values of the total number of students with disabilities registered at the University of Kent

	2007–2008		2008–2009		2009–2010	
Categories of disabilities	Number	%	Number	%	Number	%
Blind/partially sighted	36	2.9	42	2.8	42	2.8
Deaf/hearing impaired	58	4.7	48	3.2	48	3.2
Wheelchair users/mobility difficulties	61	5.0	48	3.2	48	3.2
Personal care needs	1	0.1	1		1	
Mental health difficulties	61	5.0	126	8.5	126	8.5
Unseen disabilities	233	19.0	230	15.6	230	15.6
Multiple disabilities	45	3.7	46	3.1	46	3.1
Autistic spectrum disorders	33	2.7	56	3.8	56	3.8
Specific learning difficulties, e.g. dyslexia	566	46.1	686	46.4	686	46.4
Disability not listed above	135	11.0	176	11.9	176	11.9
Information refused	1	0.1	14	0.9	14	0.9

Adopting the inclusive design principles, the auditor studied how well each feature satisfied the accessibility requirement for the users listed in Sect. 5.5.1.2.

Tables 5.2 and 5.3 present an example of the population distribution at the University of Kent and the disability types and statistics.

Figure 5.3 illustrates the results of the online survey, showing the distribution of disabilities at the University of Kent, Canterbury campus.

The access audit report provided detailed descriptions of the features of each building, with suggestions for improving access, and the summary highlighted examples of good and bad practice with respect to each feature. The features were rated in accordance with the criteria displayed in Table 4.1.

Table 5.3 Number of students by disability classification with percentage values of the total number of students with disabilities registered at the University of Kent from 2010 till 2012

Categories of disabilities	2010–2011		2011–2012	
	Number	%	Number	%
Blind/visual impaired	37	2.3	31	1.7
Deaf/hearing impaired	43	2.6	33	1.9
Physical/mobility difficulties	60	3.7	74	4.3
Personal care needs	1		0	
Mental health difficulties	189	11.7	270	15.4
Unseen disabilities	239	14.6	213	12.0
Multiple disabilities	65	4.0	104	5.8
Autistic spectrum disorders	50	3.0	63	3.7
Specific learning difficulties, e.g. dyslexia	736	45.0	785	44.2
Disability not listed above	210	12.9	196	11.0
Information refused	4	0.2	0	

Types Of Disability

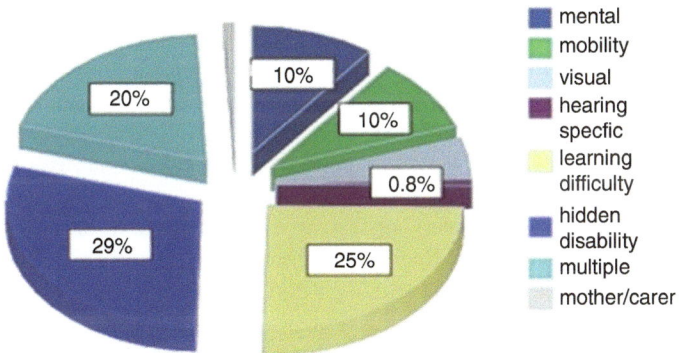

Fig. 5.3 Results of the online survey of types of disability

5.5.1.2 Access Audit Results

The access audit results answered the following research questions:

1. What are the physical and management barriers in buildings which hinder individuals with disabilities from accessing them?
2. To what extent does the University of Kent comply with the Building Regulations to ensure access and eliminate architectural barriers?

The results are thematically presented. Two aspects were covered by the access audits: (1) physical environment containing results of external and internal environment features and (2) management procedures and practices.

Table 5.4 University bus drop-off zone

Bus drop-off zone	Buildings
Keynes College	Venue Jarman Building Eliot College Marlowe Building
Darwin College	Marlowe Building Templeman Library Registry

External Physical Environment Results

Results concerned with the external environment of all six buildings cover the following physical features: bus stops, parking spaces and accessible bays, pedestrian routes, and external steps and ramps.

Bus Drop-Off Zone

The university provided a bus drop-off zone that students could use to reach their desired destination. Table 5.4 details the accessed bus drop-off zone.

The Jarman Building, the Venue, Eliot College and the Marlowe Building could be reached by using the Keynes College bus drop-off zone. The pedestrian route that operates from University Road and Keynes College bus drop-off zone was accessible to all potential users, with its dropped kerbs and tactile warnings that assist people with visual impairments to detect the approach of the dropped kerb. In contrast, the pedestrian route serving Darwin College bus drop-off zone and surrounding buildings, such as the Registry and Darwin College, did not have dropped kerbs or blister tactile warnings. Although Part M does not cover the route to a building, identifying barriers within this route was important in order to evaluate the total journey users undertook to reach designated buildings. Figure 5.4 illustrates the absence of dropped kerbs and tactile warnings at Darwin College drop-off zone.

Access auditing of the timetables that were posted at the two drop-off zones revealed that not all buses provided low floors, which resulted in delay for some students with disabilities, preventing them from reaching university at their desired time.

Car Parking

Whilst Eliot College, the Templeman Library and the Registry buildings had their own parking spaces that provided accessible parking bays, the Marlowe Building, the Jarman Building and the Venue did not have their own designated parking spaces. Accordingly students and staff members with disabilities using cars to reach the Marlowe and Jarman Buildings and the Venue had to park their cars in Eliot College's two accessible parking bays, which could not satisfy the demand of all users with disabilities. Table 5.5 summarises the car parking spaces available and the main accessibility issues at each building.

Fig. 5.4 Absence of dropped kerbs and tactile warnings along pedestrian route leading to Darwin College bus drop-off zone

Table 5.5 Accessible car parking bays at six audited buildings

Car parking	Building	Accessible bays numbers	Accessibility issues (features that do not comply with Part M)
Eliot College	Eliot College Venue Jarman Marlowe	2	• No signage indicating bays • No hatched access transfer for boot access • Unclear markings of bays • Loose surfaces
Registry	Registry	3	• Kerbs between the parking bays and routes did not have tactile warnings
Templeman Library	Templeman Library	2	• No signage indicating bays • No tactile warning indicating dropped kerbs • No hatched access transfer for boot access • Not enough space for car doors to be fully opened to allow drivers to get in or out to transfer • Kerbs between the parking areas not dropped • No tactile warnings provided

The key features that did not comply with Part M Building Regulation were identified when access auditing the accessible parking bays and were as follows: the absence of signage indicating parking and the absence of a safety-hatched transfer zone to allow boot access, poorly maintained surfaces (Fig. 5.5a), and absence of dropped kerbs and tactile warnings (Fig. 5.5b).

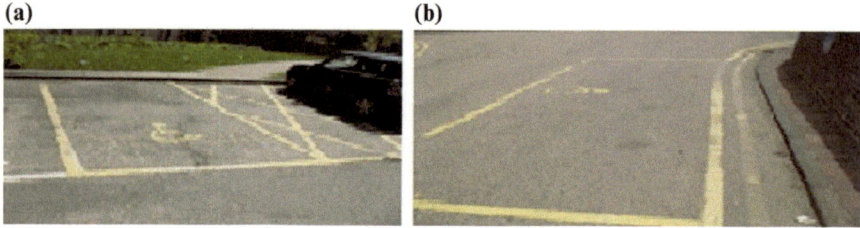

Fig. 5.5 Accessible car parking bays at (**a**) Eliot College and (**b**) Templeman Library

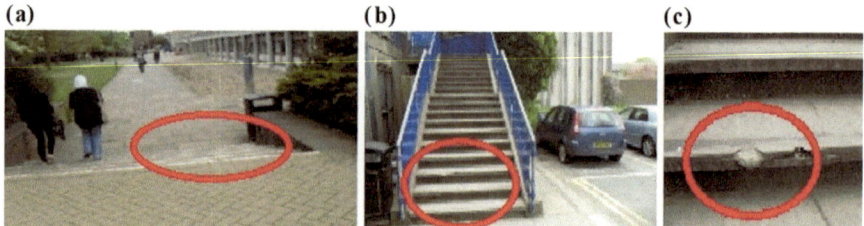

Fig. 5.6 External steps. (**a**) Marlowe Building, (**b**) Eliot College, (**c**) Eliot College

External Steps

Eliot College and the Marlowe building had external steps on the pedestrian approach route. The fundamental accessibility barriers that did not adhere to Part M standards were the absence of corduroy tactile warnings at the top and bottom of steps, lack of contrasting nosings on steps, absence of handrails along steps (Fig. 5.6a) or handrails that did not extend horizontally at the top and bottom of steps (Fig. 5.6b), and uneven risers and poorly maintained step surfaces that could present trip hazards and obstruction (Fig. 5.6c).

External Ramps

External ramps were provided in the Marlowe Building, Eliot College, the Registry and the Templeman Library. Whilst the ramp provided at the Registry was a good example of an accessible ramp, with adequate gradient, width, landing and handrails, which had visual contrast against its background (Fig. 5.7a), the ramps provided at the Marlowe Building, Eliot College and the Templeman Library did not follow the ramp design guidelines stated in the Part M regulations and hence they presented many barriers that limited accessibility for people with mobility and visual impairments.

The ramp provided near the Marlowe Building and the Templeman Library had a steep gradient, uneven surface finishes and absence of handrails along the ramp (Fig. 5.7b, c), whilst the ramp provided at Eliot College had a long ramp with a steep gradient and narrow width, in addition to the absence of a kerb on the outer edge, a difference in frictional floor level, and handrails placed at a high level, which did not comply with standards (Fig. 5.7d).

Fig. 5.7 External ramps. (**a**) Registry, (**b**) Marlowe Building, (**c**) Templeman Library, (**d**) Eliot College

Internal Physical Environment Results

The audit of the internal environment of the six buildings covered the main entrance, the reception counter, corridors, lifts, stairs, ramps, toilets and emergency egress routes.

Main Entrance

The level of accessibility at the main entrances varied. Many main entrances did not comply with Part M regulations and were problematic for a variety of users. The key problems identified were the absence of, or inadequate, signage, level access, canopies over entrances, inadequate contrasting strip manifestations on glazed doors (Fig. 5.8a, b), lack of automatic or power-assisted doors, and firm and flush entrance mats.

The Jarman Building School of Arts did not have a sign to indicate its name outside the main entrance, whilst the sign that was placed inside the internal lobby was at such a high level that it was hard to see it. In addition the glazed entrance doors had two frosted strip markings or manifestations that were not clearly visible against a variety of backgrounds and in different lighting conditions (Fig. 5.8a). In a similar case, at the Registry, the strip door manifestation did not have adequate contrast (Fig. 5.8b). The Registry, the Venue, the Marlowe Building, Eliot College and the Templeman Library had the same external signage design that lacked adequate visual contrast with the background (Fig. 5.7b–f).

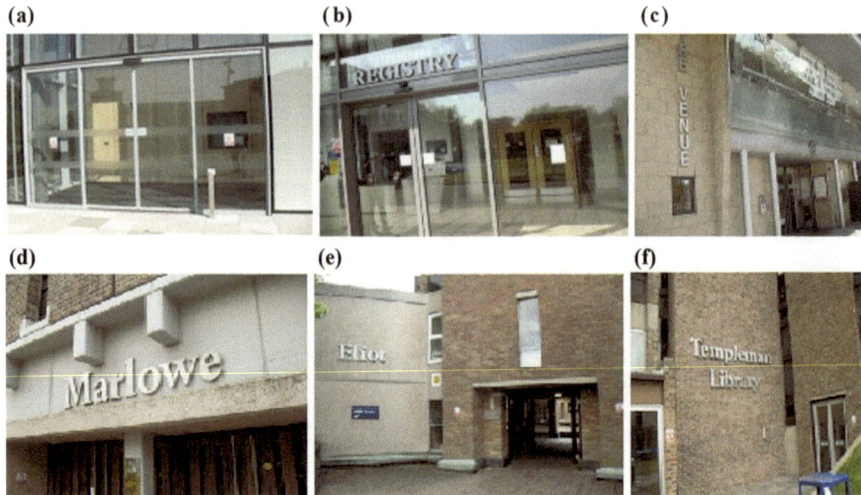

Fig. 5.8 Main entrances. (**a**) Jarman, (**b**) Registry, (**c**) Venue, (**d**) Marlowe Building, (**e**) Eliot College, (**f**) Templeman Library

Whilst the Jarman Building and the Registry had automatic entrance doors that provided adequate accessibility for all users (Fig. 5.7a, b), the Marlowe Building and the Templeman Library provided power-assisted main entrance doors in addition to manually operated doors that were heavy to open. Similarly, the Venue had main entrance doors that were heavy to open. Eliot College had a sliding door that was permanently held open.

Reception Counter

Reception counters tended to be poorly designed and failed to comply with Part M regulations. There was a lack of low counters at the Venue and Eliot College; no induction loops at the Registry, the Marlowe Building, the Venue and Eliot College; and no knee recess sections and insufficient manoeuvring space at the Registry, the Marlowe Building and Eliot College. Whilst some buildings provided a low counter for wheelchair users, for example the Jarman Building, either the counters were inappropriately positioned, so wheelchair users did not know that they existed (Fig. 5.9a); or the needs of those with hearing impairment were neglected (lack of induction loops, Fig. 5.9b); or there was an absence of signage even when induction loops were provided (Fig. 5.9a).

Corridors

The corridors and passageways inside the six buildings differed in terms of their width and internal surfaces and did not all convey directional information about the building or assist with circulation around it (Fig. 5.10a). With the absence of

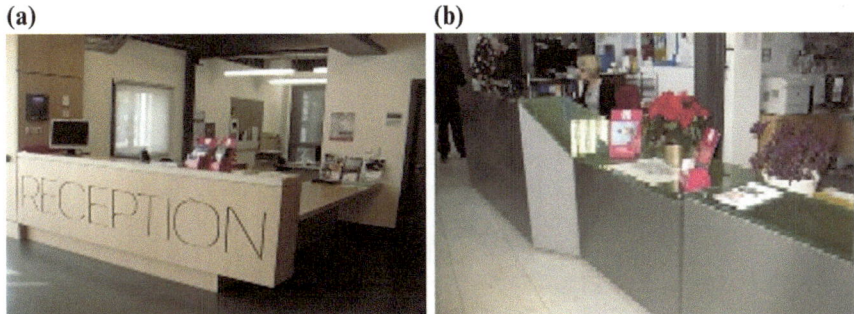

Fig. 5.9 Reception counters: (**a**) Jarman Building, (**b**) Registry

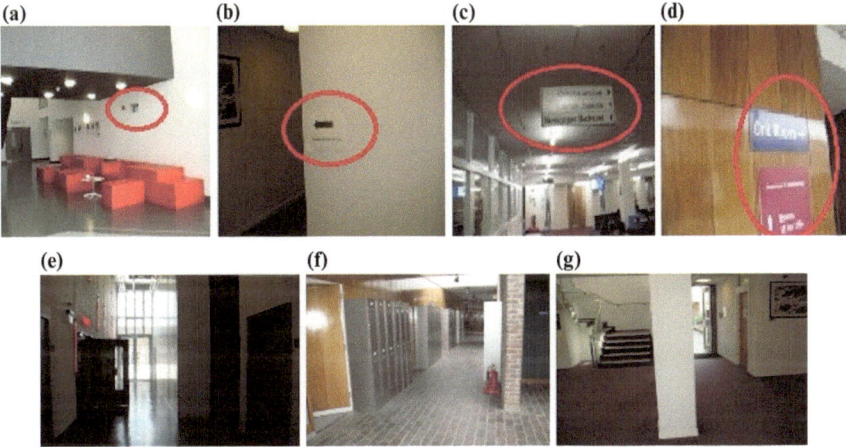

Fig. 5.10 Corridors and signage: (**a**) Jarman Building, (**b**) Registry, (**c**) Templeman Library, (**d**) Marlowe Building, (**e**) Jarman Building, (**f**) Marlowe Building, (**g**) Registry

directional signs, first-time visitors and others may find it difficult to reach their designated space or office; hence directional signage plays a vital role in wayfinding and getting around buildings. Where directional signage was provided in corridors, it was either temporary or paper based in nature that can be easily torn or fall on the ground limiting people to find their way to a designated facility (Fig. 5.10b). Some of the directional signs had arrows that were misleading people following them as they were put in a confusing position. In addition, some of the lighting features were placed closely to the signage, creating reflections that may hinder many people from reading the signage directions and information (Fig. 5.10c). Moreover, many building signs had not been updated to display the new name of the room, such as the Crit Room/the M Arch room (Fig. 5.10d).

Passenger Lifts and Platform Lifts

Also, some corridors had doors that opened outwards, which limited the space for manoeuvring and created obstruction (Fig. 5.10e); other doors were not easily distinguishable from the corridor finish (Fig. 5.10f). In addition, the Registry building had a column along the passageway route that did not have a band of contrasting colour to indicate its presence to people with visual impairment (Fig. 5.10g).

Passenger lifts were provided in the six buildings; however, they were not clearly signposted to indicate their locations. Whilst these lifts were designed to provide access for wheelchair users, they were not well designed for such individuals or those with visual or cognitive impairments: call buttons were not within the reach of a wheelchair user, and there was poor visual contrast, poor lighting and an absence of audible announcements, Braille characters or embossed buttons, and horizontal handrails. Some lifts did not provide access to all floors, in which case a platform lift had been installed. Table 5.6 summarises the main accessibility issues covering the passenger lifts provided in the six buildings.

Platform lifts were provided at Eliot College and the Templeman Library. Whilst the platform lift at the Templeman Library was used to provide access to Level 1, its internal dimensions could only accommodate one wheelchair user. The platform lifts at the Templeman Library and Eliot College did not have a mirror to aid in reversing out of the lift, neither did they have a visual floor indicator. In addition, the platform lift at Eliot College did not have an audible announcer, Braille or tactile markings. Although it had an emergency button, it was not clearly signposted; there was merely a handwritten sign that read 'only use stop push in an emergency' (Fig. 5.11).

Internal Staircases

Although access to different levels inside the six buildings could be achieved via steps and lifts, some staircases were poorly designed in certain respects; for example, they had open risers and a block wall in front of a landing (Fig. 5.12a); handrails

Table 5.6 Passenger lifts

Lift accessibility issues (features that did not comply with Part M)									
Buildings	No directional signs	Do not reach all floors	No Braille characters	Unreachable buttons	No handrail	No audible sound	Dim lighting	No tactile marking
Jarman	x		x	x				
Venue	x				x			
Eliot	x			x	x			
Templeman		x		x	x	x	x	
Marlowe	x							
Registry	x		x		x			x

Fig. 5.11 Eliot College platform lift interior and signage

(a)   **(b)**

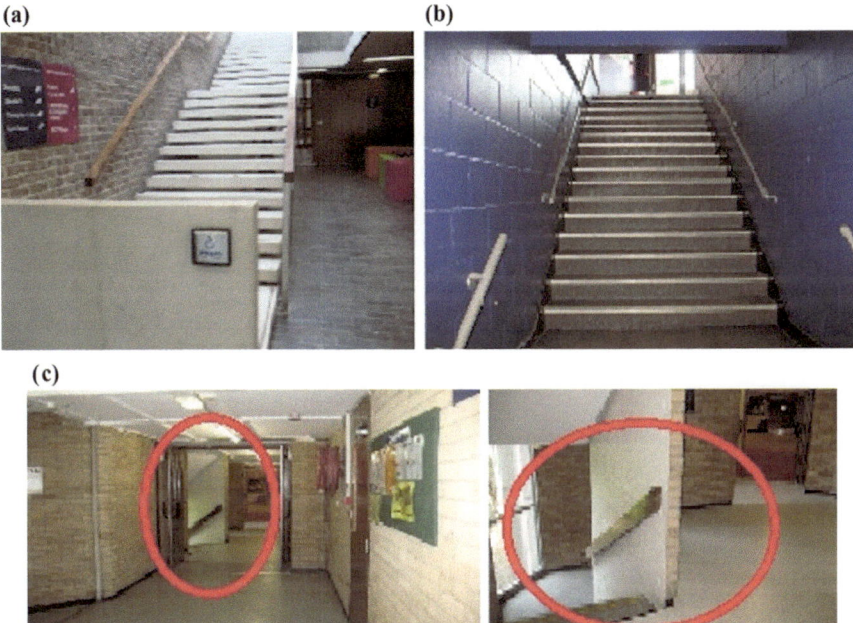

(c)

Fig. 5.12 Internal staircases. (**a**) Marlowe Building staircase. (**b**) The Venue staircase. (**c**) Eliot College West Wing staircase

that did not extend beyond the top and bottom of steps or run along landings (Fig. 5.12b); and a lack of clear contrast with adjacent surfaces and backgrounds. In addition, some staircases were difficult to locate as they were not immediately visible from the building concourse, and there was an absence of signage directing users to them (Fig. 5.12c).

Toilet Compartments

Toilets for male, female and unisex wheelchair users were provided at the six buildings.

Female and Male Compartments

Whilst female and male toilet compartments were designed to meet the needs of able-bodied users, they were not well designed for people with visual and mobility impairments. The major barriers identified were absence of Braille or embossed signage on doors, heavy doors that were hard to open, lack of space for ambulant users in both female and male cubicles, lack of grab rails and coat hooks. In addition WC seats, urinals and sanitary units where provided are inappropriately placed either too high to be reached or too low. Another key issue was basins with separate knob taps that were difficult to operate for people with limited dexterity. Poor visual contrast of fixtures and surface finishes and poor lighting were other common issues identified in both female and male toilets. Figure 5.13a gives an example of signage used in the Jarman Building that was problematic for people with visual impairments. The pictograms with a shiny and reflective finish were mounted on white walls and created glare for people with visual impairment.

In addition, Fig. 5.13b shows urinals provided at Eliot College placed at a low level, whilst Fig. 5.13c gives a good example of bad design in terms of poor visual contrast between floor, walls and WC seat; inappropriately placed grab rails; and separate knob taps.

(a)

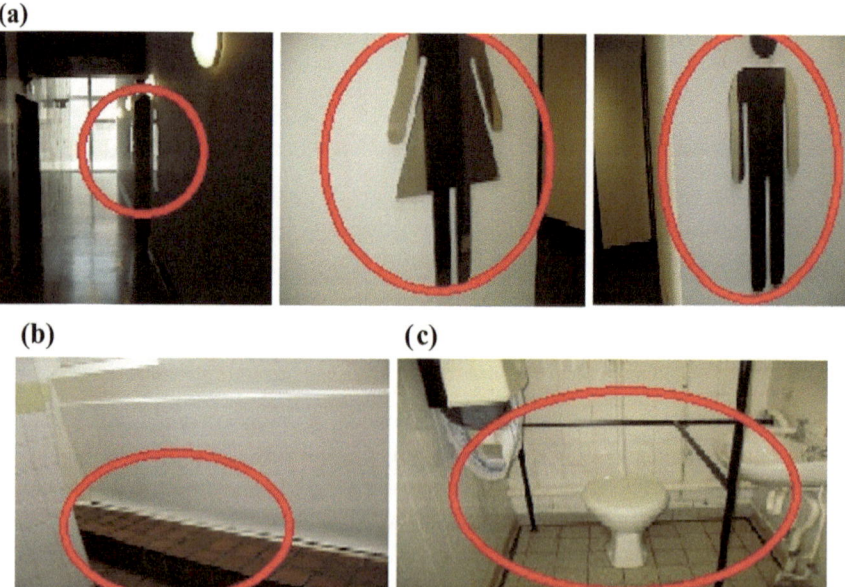

(b)   **(c)**

Fig. 5.13 Toilets. (**a**) Jarman Building toilet signage along corridor. (**b**) Eliot College male toilet. (**c**) Templeman Library toilet

Unisex Wheelchair-Accessible Toilets

Wheelchair-accessible toilets were provided in all six buildings; however, there was a lack of directional signage to indicate their locations. Although accessible toilets were designed to meet the needs of wheelchair users in the first place, there were many accessibility issues affecting wheelchair users and also those with visual and cognitive impairments. The absence of Braille or embossed signage could prevent a person with visual impairment from discovering the location of the toilet. Moreover, heavy doors with no visual contrast; inappropriately positioned door handles, grab rails, WC seats and sanitary fittings; and poor visual contrast of fixtures and fittings could also limit many users from accessing the toilet. Where an emergency alarm was provided, it was out of reach. In addition, lack of horizontal closing bars on the inside face of doors, absence of shelves and coat hooks and poor maintenance contributed to the inaccessibility of these facilities. Table 5.7 illustrates the accessibility issues identified in the unisex toilets provided in the six buildings.

At the Venue and Eliot College two accessible toilets were provided but these were difficult to locate because of the absence of directional signage. In both cases, the rooms were of a suitable size but defects included poor contrast of fixtures and fittings; poor maintenance, namely broken grab rails (Fig. 5.14a); ER emergency alarm out of reach; and separate knob taps (Fig. 5.14b).

Table 5.7 Barriers to unisex wheelchair-accessible toilets in the six buildings

Buildings	No Braille signs	Heavy doors / No horizontal closing bar	No visual contrast in doors / Fixtures no visual contrast	High Sanitary units	Low WC seat / High/low Grab rails	ER alarm out of reach / Manual lighting	No shelf / Coat hook	
Jarman	X	X				X		
		X			X			
Venue	X	X	X		X	X		
		X	X		X	X	X	
Eliot	X	X		X	X	X	X	X
		X	X		X	X	X	
Templeman	X	X	X	X	X	X	X	
			X		X	X	X	
Marlowe	X	X	X	X	X	X		
			X		X			
Registry	X			X	X		X	
		X	X		X	X	X	

(a)   **(b)**

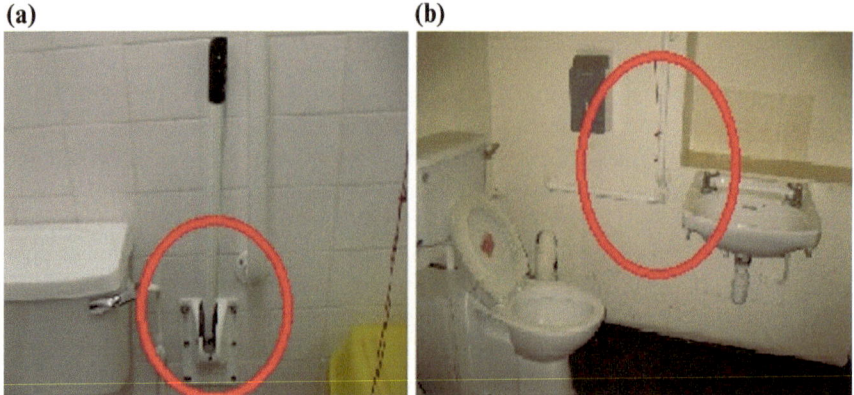

Fig. 5.14 Accessible toilets. (**a**) Eliot College. (**b**) The Venue

Emergency Egress Routes

Emergency egress routes were provided in all the six buildings. Although these emergency exit routes were clearly signposted throughout the buildings, there were some issues limiting accessibility. Although glazed doors were used as emergency exit routes, there were no strip door manifestations that contrasted with the variety of backgrounds. In addition, emergency doors that were manually operated with push pads were difficult to operate for people with limited dexterity (Fig. 5.15a). Also, fire extinguishers placed near the egress route and unguarded could cause an obstruction for people with visual impairments (Fig. 5.15b). Where staircases were used as an emergency exit route, steps did not have any nosings, and handrails did not extend to the top or bottom of steps and did not run along landings (Fig. 5.15c).

Access Statement for the Jarman Building

Analysis of the access statement for the Jarman Building revealed that although it emphasised applying inclusive design to the external and internal features of the building, many of these recommendations had not adopted in the implementation phase.

Site Approach and Public Transport

The statement emphasised a new provision that was to be created on University Road for drop-off/collection immediately adjacent to the new building; however, the access audit revealed that such provision had not been introduced.

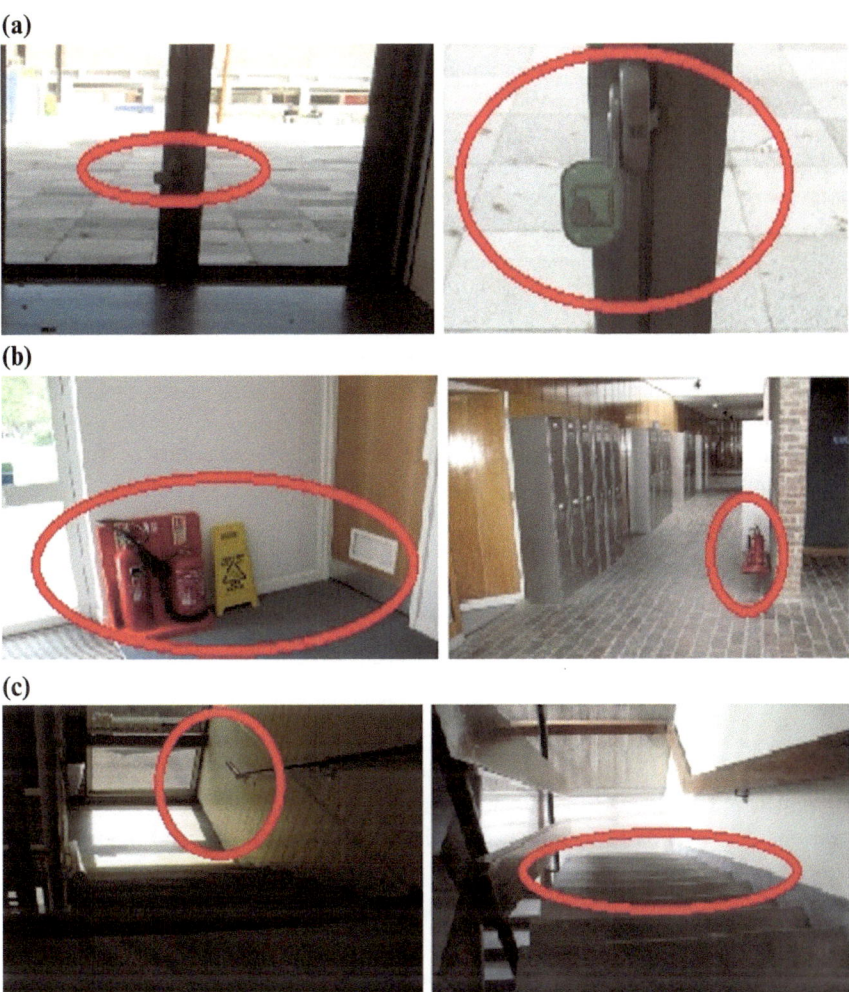

Fig. 5.15 Emergency exits: (**a**) Jarman Building—glazed emergency exit door with no strip door manifestation and a manually operated push-pad handle, which is hard to operate for people with limited dexterity. (**b**) Registry and Marlowe Buildings with fire extinguishers placed near exit route. (**c**) Eliot College exit staircase with handrails that do not extend to the top and bottom of steps and Marlowe Building exit staircase with steps without nosings

Arriving

The statement placed emphasis on providing accessible parking bays and car drop-off zones near the Jarman Building. To respond to the access statement recommendations, designated accessible parking bays had been provided at the rear of the building; however, car drop-off zones and visitors' parking bays were not provided despite having been as suggested in the statement.

Entrance

The statement proposed installing an automatic door accessible to all users. It also proposed that a warning strip or safety logo should be placed at two heights, compatible with the eye level of adults and children, respectively. Although there was an automatic glazed door with frosted warning stripes at the main entrance, these stripes were not effective because they did not provide adequate visual contrast for users with visual impairments.

Lifts

Although the statement emphasised the need for a passenger lift that could accommodate a wheelchair user, it did not specify the needs of other users, that is, those with visual and hearing impairments. The access audit revealed that the passenger lift in the Jarman Building did not comply with the provisions of Approved Document M.

Circulation

The statement emphasised the need to provide doors that were light enough to be used by people with limited mobility or strength, but most access audited doors were too heavy to be opened easily.

Stairs

Compliance with the access statement was noted in the case of staircase provision in terms of provision of nosing and handrails that extended beyond the steps and onto the landing.

Toilets and Showers

The statement proposed unisex accessible toilets, showers and cubicles for ambulant users in standard toilets that would comply with BS 8300:2018 and Approved Document M. Whilst these provisions were met inside the building, many of the fixtures did not comply with these standards.

Signage and Navigation

The access statement proposed clear directional and information signs in compliance with BS 8300:1, 2 2018 to meet the needs of all users including people with visual impairments and learning difficulties. However, the access audit revealed that the building failed to implement the access statement recommendations.

Lighting and Decor

The access statement proposed the use of glare-control measures using blinds to reduce reflection. It also proposed the use of colour coding as a means of assisting orientation on each floor. However, such recommendations had not been adopted inside the building.

Analysis of the access statement highlighted the fact that although the required inclusive design features had been written into the statement in order to obtain approval at both planning and Building Regulation application stages, many of these inclusive features had not been provided by the architects and executive team in the construction and implementation phases at the Jarman Building.

Management Practices and Procedures

There were few management practices and maintenance issues identified that affected accessibility. Those that were found included the stacking of office furniture and stationery in spaces that limited the manoeuvring space for wheelchair users and created an obstruction for people with visual impairments (Fig. 5.16a). Another example of bad management was (Fig. 5.16b) a help-point push-button in the refuge area being blocked by two bins, which made it impossible for a wheelchair user to use in case of emergency and could cause an obstruction for people with visual impairments. Moreover, leaving scaffolding in front of a main entrance (Fig. 5.16c) and placing movable book carriers and shelves along the route could limit the manoeuvring space for wheelchair users and cause a trip hazard for people with visual impairment.

5.5.2   Consultation with Students and Staff Members with Disabilities

5.5.2.1   Overview

The purpose of the consultation with students and staff members with disabilities was to obtain feedback with respect to the barriers they encountered when accessing the external and internal environment of the university buildings. The author conducted ten personal interviews with students and staff members with disabilities. The interviewees had disabilities which varied from mobility difficulties, special learning needs, visual impairments, mental health difficulty and dyslexia to multiple disabilities.

Seven interviewees had had a disability from birth; three of them had become disabled after birth but before joining the university. In addition, 236 participants, namely 182 undergraduates, 36 postgraduates and 17 staff members, completed the online questionnaire. The survey was circulated among all disabled students registered at the

(a)

(b)

(c)

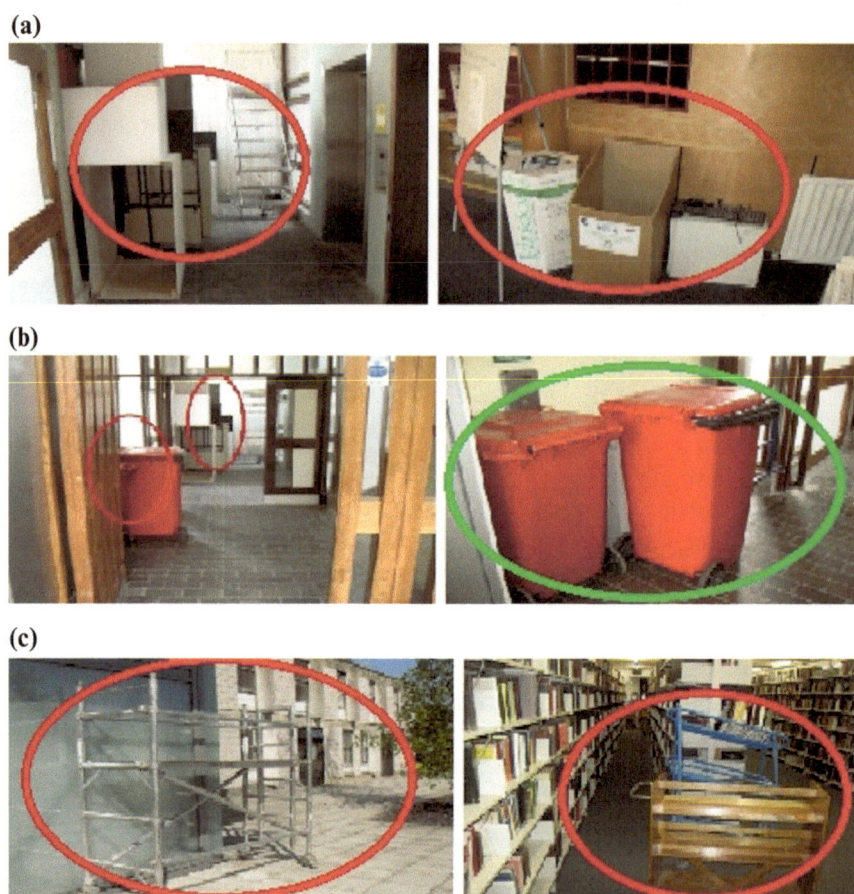

Fig. 5.16 Bad management practices. (**a**) Bad management practices in the Marlowe Building and Eliot College. (**b**) Bad management practices in the Marlowe Building. (**c**) Bad management practices in the Registry and the Templeman Library

Disability and Dyslexia Student Services (DDSS) and a representative sample of 1000 non-disabled individuals. The online survey at the University of Kent revealed that 33% of the population who replied to the questionnaire were disabled, 18% were partially disabled (hidden or temporary disability) and 49% were non-disabled (Fig. 5.17). Moreover around 67% of the population were females, and 33% were males, and 80% of them were British, whilst 20% were international. The surveyed population was categorised in terms of three different age groups: 84% were between 18 and 35 years old, 9% were between 35 and 49 years old and 7% were between 50 and 65 years old.

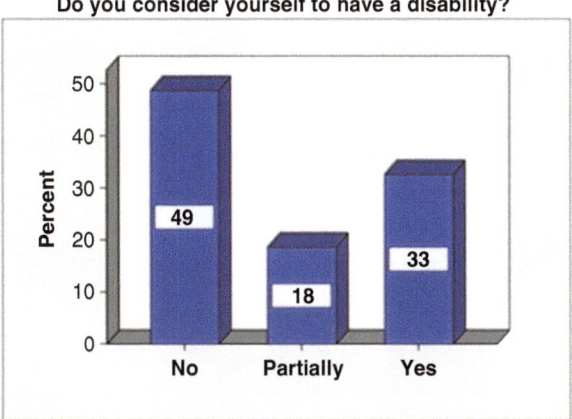

Fig. 5.17 Online survey: statistical data about participants with disabilities and non-disabled at the Canterbury campus

5.5.2.2   Results

This section presents answers to the following research questions:

1. To what extent are students, with and without disabilities, and staff members satisfied with the services provided at the University of Kent?
2. Which inclusive/accessible approach would users most like to see adopted at the University of Kent, Canterbury?

The qualitative and quantitative data are thematically presented, with seven sections combining the participants' input from interviews and the online questionnaire responses. The thematic sections contain information about (1) external physical environment barriers; (2) internal physical environment barriers; (3) management procedures and practices; (4) staff training and disability awareness; (5) effectiveness of legislation; (6) integration and consultation with disabled people; and (7) inclusive approach: design, policies, practices and procedures.

External Physical Environment Barriers

Participants' responses showed that public transport, buses and cars were the main transport means used by the majority of users. Figure 5.18 shows a breakdown of the means of transport used. Users with no disabilities favoured public transport and their own cars, whilst participants who were partially disabled and those with disabilities used their own cars as the primary means of transport.

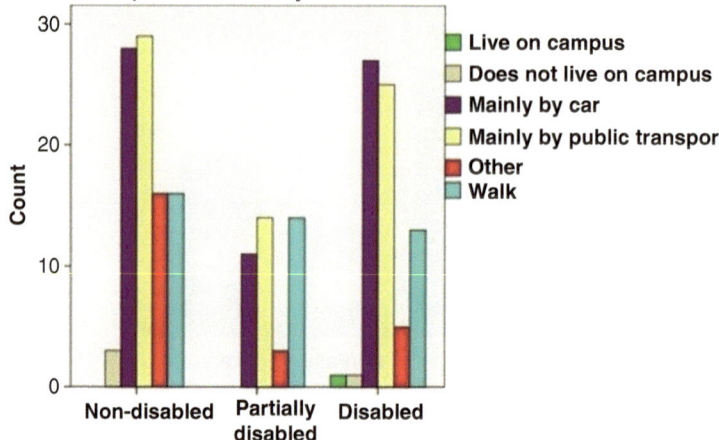

Fig. 5.18 Means of transport used to travel to Canterbury campus

Accessible Parking Spaces

Participants' responses revealed that the number of parking spaces for users with disabilities was insufficient, suggesting that the university should provide more accessible parking bays. Figure 5.19 shows the importance of providing accessible parking bays: around 39% of participants stated that such bays were very important when it came to enhancing accessibility and promoting integration.

In addition, when a car parking space was available and was close to the main entrance, either it was positioned on a slope or it did not have sufficiently dropped kerbs and it was necessary to walk a long way to find a level route. One student noted:

> Sometimes they put accessible parking bays nearer to building but an gradient which is uphill, so then I get the wheelchair up and because I have the wheelchair next to me, bringing up the car, and then trying to get the wheelchair on, assemble it and at that point go downhill, it is hard to manage all of that with such a gradient!

One student pointed out that car parking did not provide a drop-off and pickup point to enable parents/carers to drop students at a convenient place.

> I wish there was a car parking for visitors, because my parents drop me at the university and university parking is at the back and you cannot get further, and people use the accessible car parking. That is the main issue.

On the other hand another participant who was both a student and a staff member pointed out that although she did not have any issues with car parking, people without a blue badge often tended to park in accessible bays.

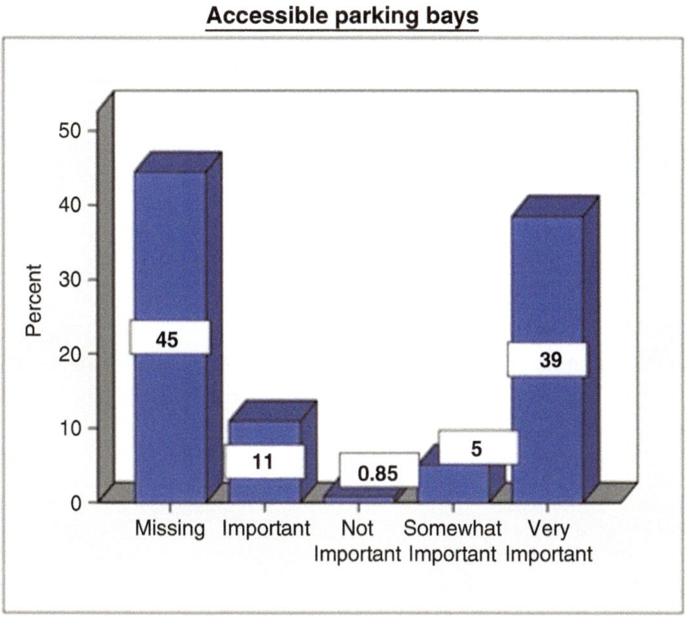

Accessible parking bays

Fig. 5.19   Importance of accessible parking bays

Buses

Only two students reported that they used the university bus. Although one student found the experience pleasant and straightforward, the second noted that not all buses had a flat or level floor, which she found hard to manage when using crutches.

External Steps

External steps were evaluated by participants with visual and cognitive impairments, whilst participants with mobility impairments summarised their experience of using external ramps. The major barriers when using external steps were (1) steep and long flights of stairs; (2) inconsistency in depth of steps; (3) lack of nosing or loose nosing; (4) lack of tactile warnings; and (5) handrails with slippery surfaces.

External Ramps

Participants with mobility impairments noted that ramps were very long and steep, and thus difficult to manage independently. 'When I use the manual chair, somebody has to push me all along the long ramp and the gradient is too much to manage by myself without having assistance'. Another participant noted that the surfaces, entrance and exit angles and the way the ramp met the door were all critical for people with mobility impairments. Whilst handrails were provided, many participants mentioned that they did not provide enough support because either they were too high or they had slippery surfaces, especially in adverse weather. Two participants mentioned that they had hurt themselves whilst accessing the ramps because of steep gradients and uneven surfaces, and noted especially that drainage channels were not flush with ramp surfaces. 'I broke my leg on this campus when the chair did tip forward underneath because of drainage barriers'.

Internal Barriers

Participants' responses showed that level access and automatic doors were two of the main factors helping to create an accessible environment. Figure 5.20 shows the importance of these two features.

Main Entrance

Participants considered that although some main entrances were more accessible than others because they had automatic doors, many still needed improvement. One participant with visual impairment mentioned that glazed doors presented a problem for her because of the lack of door manifestations. 'Glass doors present a bit of a problem to me when they are very clean. In Keynes I cannot tell where the door is; you risk walking into doors so it is hard to know the way'.

Another student mentioned that visibility of glazed doors could be improved by using clearly visible and distinguishable edges. All participants mentioned that heavy doors, inadequate placement of handles and inappropriate shape of handles hindered them from accessing entrances independently. 'Some doors are very narrow, and some doors are too heavy, and you cannot get in and there is no way to get in or out'. Another participant believed that the shape and position of door handles above the floor were the main features that restricted users from opening the doors.

> As for handles, they are always too high or the shape is wrong in two ways. It is difficult to grasp. Also about handles and wheelchair users, I find the doors have slightly recessed glass so when you are driving a wheelchair or power wheelchair, the handle catches the chair or catches you on the shoulder. I always go with a lever handle you can operate with any part of your body, elbow, nose etc., as for a knob, you have to have a good grip.

When asked if they favoured automatic doors, power-assisted doors or manually operated doors, all participants opted for automatic doors. 'Automatic doors are easier, they open and you don't have to worry about finding handles'.

One participant showed his concern about relying on high-tech solutions, such as automatic and power-assisted doors, which needed regular maintenance.

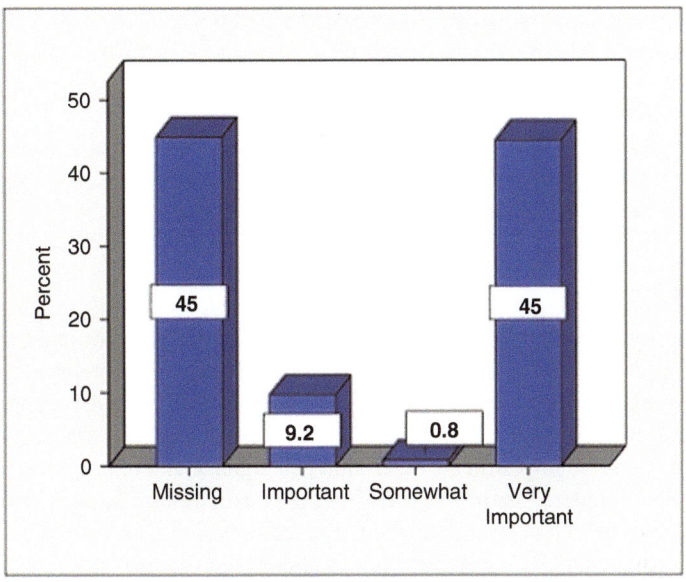

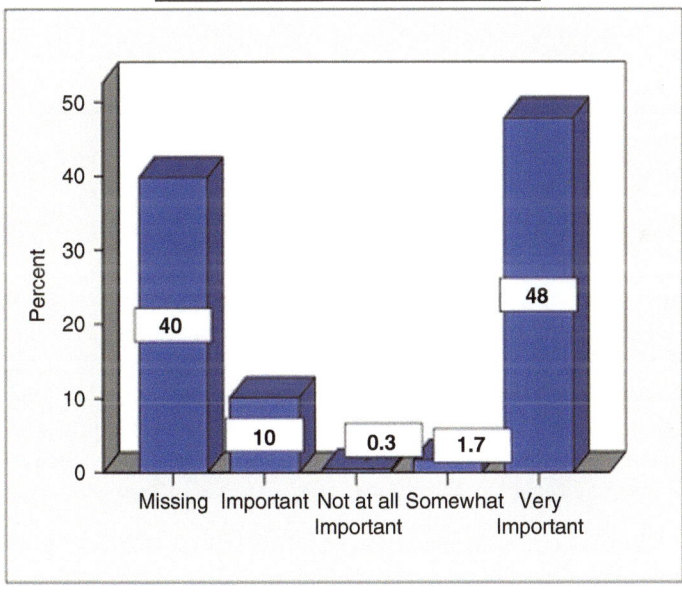

Fig. 5.20 Importance of level access and automatic doors

Counters

Among the participants with mobility impairment who were interviewed, five out of ten had experienced difficulty in approaching counters because of their height. 'The main problem with the counter desk is the main space inside, and they all don't have a lower counter'.

Also, participants with visual and cognitive impairments pointed out that the maps provided at reception counters were difficult to read because of the small print and lack of tactile or Braille-marked maps.

Corridors

Corridors were problematic for all interviewed participants, whether they had mobility, visual, cognitive or hidden impairments. The main problems encountered were poor signage, dim lighting, inappropriate corridor width, heavy doors that open outwards, inappropriate floor finishes and lack of handrails. Participants with visual and cognitive impairments noted that bad signage and dim lighting were the barriers which hindered them from reaching the desired destination. Three participants with visual impairment described signage provided in corridors as follows: 'Signage is diabolical ... it is awful .... Out of date and ripped down ... In a similar scenario one participant with cognitive impairment commented: I get confused because of my conditions and I panic and I get lost because signage is so confusing'.

In addition, interviewed participants with mobility impairments noted that narrow and crowded corridors limited their manoeuvring space. One participant with multiple sclerosis (MS) noted:

Corridors have open windows which make the environment cold and this affects people with MS who can be moved off the balance, so that is not ideal ...

There is lot of people coming the other way, and that is a mass effect, and it is quite over-whelming, and especially when I am walking and there is lack of awareness from people, I suppose.

Another participant with mobility impairment noted that handrails along a corridor could provide additional support: 'There are not many handrails along corridors, and it would be useful to have them, especially between the lectures, when it is crowded and sometimes you are pushed, so having handrails along the corridor will give additional support'. Two participants pointed out that the floor surfaces of corridors either were slippery or had carpet which was hard to manage for wheelchair users, and could cause an allergic reaction in people with respiratory conditions.

Internal Staircases and Lifts

Participants' online responses showed that lifts and handrails alongside steps were important for improving accessibility.

Figure 5.21 illustrates the importance of providing these two features.

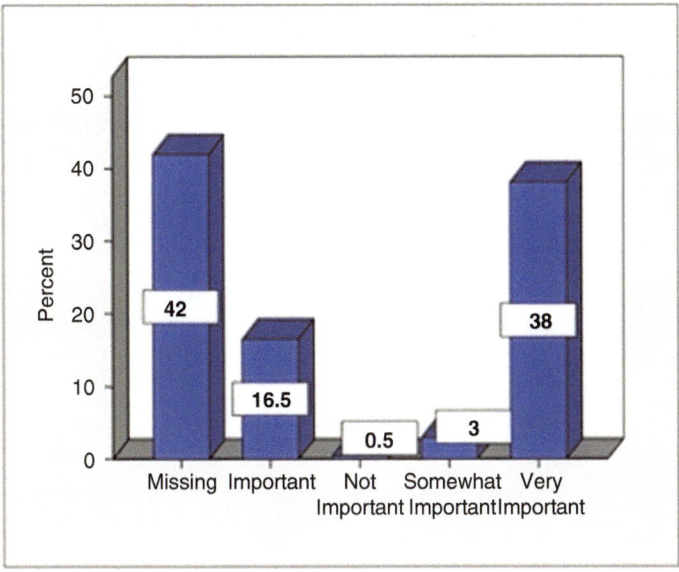

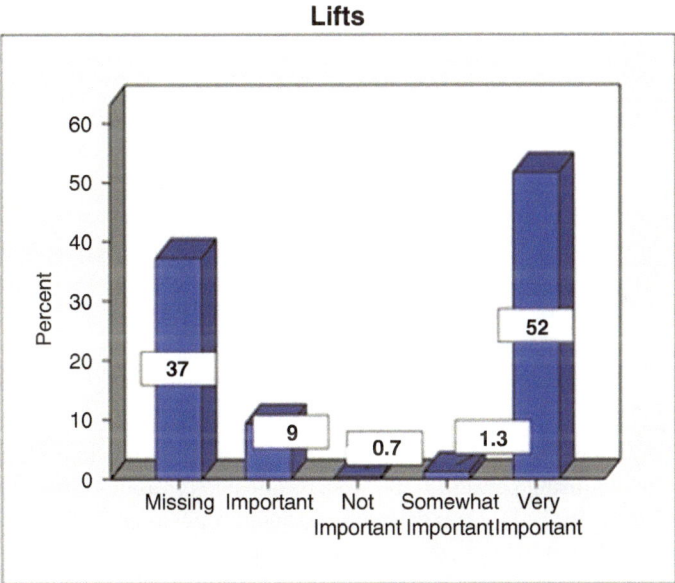

Fig. 5.21 Online survey: Importance of handrails and lifts

Staircases

For vertical circulation, staircases were used by ambulant users and those with visual and cognitive impairments, and lifts by wheelchair users. Participants with visual and cognitive impairments experienced difficulties in accessing narrow steps with open risers and with nosings that did not have adequate visual contrast.

> Steps with open risers in Keynes lecture theatre, and you feel you will put your feet through and you are not sure of the depth of steps; it is quite a hindrance, I don't think they are very practical and they are dangerous, and if there is a person with a cane they could put it and know it is there, but they can't know the depth.

In addition, two participants with visual impairments pointed out that circular steps like the ones in Rutherford College were hard to see and the depth of them was difficult to judge.

> Stairs in Rutherford are circular so you cannot see the end of the stairs, and you don't know how long they are so you cannot judge whether you can do it. There is no sign on them to tell you how far, and again, because it is invisible, there are people coming and they cannot see you.

> ... Rutherford seems to have steps that are bound to go up and down sometimes my vision get blurred and steps are very difficult. They have handrails but the stairway can be very busy. I think there should be a different form of access where it is possible.

Lifts

Participants with mobility impairments noted that the main issues with lifts were the following: (1) poor signposting; (2) inconvenience because they broke down frequently; (3) their promotion of segregation; (4) inappropriate placement of controls; (5) absence of mirrors and handrails; and (6) platform lifts that required the door to be pushed to open it. One participant with a mobility impairment pointed out that a lift in the Templeman Library was in front of the glazed screens in the main entrance, and could only be used by individuals with disabilities:

> The lift in the library is open, and I think everyone can see you and you are bit on display kind of thing and everyone watches you ... You are different; you cannot go the normal way because you cannot go up the upper stairs to the actual building.

Another participant with cognitive impairment noted that lifts in the Templeman Library were confusing because some were designated for use by students and others for staff members only: 'You get confused and it is not as clear as I feel it should be, signage to lift and lighting are poor ...'

Another point noted by participants with visual impairments was that call buttons needed to be bigger to be easy to operate. In addition, one participant with visual impairment recommended that emergency buttons should be in colours that distinguished them from the other call buttons. Another such participant favoured using steps rather than lifts because:

> I don't use the lifts at the university, because it is not clearly signposted where they are, and I get disorientation in buildings like Rutherford or Eliot so I feel if I use the lift, I will lose

my sense of orientation to where I was in college when I change floors. I sort of remember routes that I know and that sounds strange, so if I have to go somewhere different or new it frights me out. So I don't tend to use lifts that much.

Figure 5.22 illustrates the relative importance of passenger lifts for people with all types of disability who found lifts very important with regard to enhancing accessibility inside buildings.

Toilet Compartments

Participants' responses confirmed that providing accessible toilets was essential to individuals with disabilities. Figure 5.23 illustrates the importance of providing accessible toilet facilities since 50% of online survey responses stated that a wheelchair-accessible WC was very important.

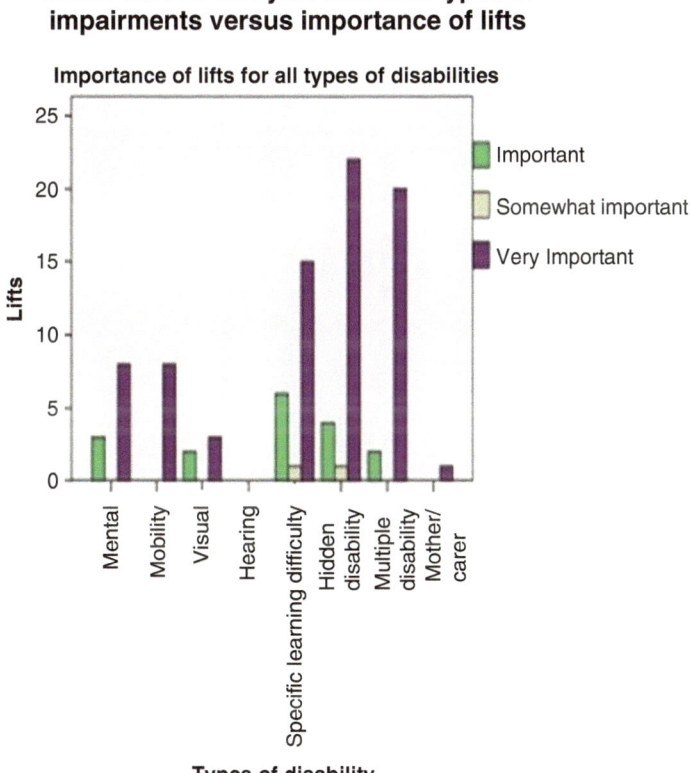

Fig. 5.22 Online survey results: importance of lifts for different impairments

Online survey at the University of Kent

Question 23
Importance of wheelchair-accessible toilet

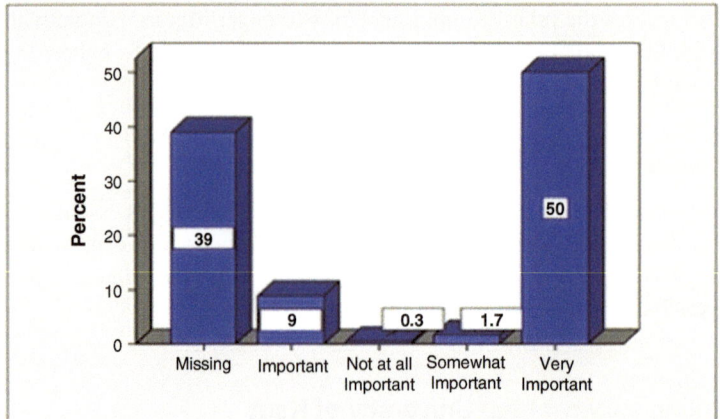

Cross.tabular analysis of importance of accessible wc & impairment types

Importance of wheelchair-accessible WC for all disabilities

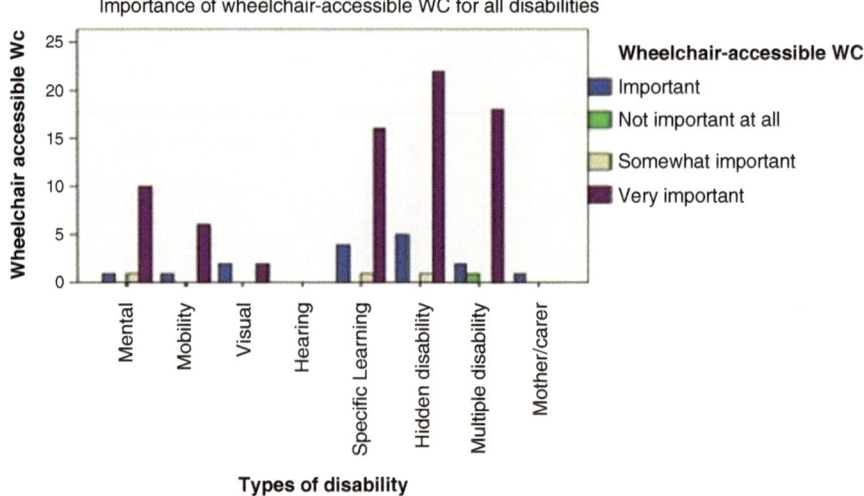

Fig. 5.23 Online survey results: importance of wheelchair-accessible toilets

Female and Male Compartments

Participants with mobility difficulties and visual and cognitive impairments expressed their concerns about using female and male toilet compartments. The main barriers were (1) lack of directional signage, (2) heavy doors, (3) lack of visual contrast, (4) absence of grab rails, (5) lack of shelves and (6) dim lighting.

One participant with mobility difficulties mentioned that she avoided using university toilets in Rutherford College and Eliot College because they were very narrow.

Another participant added that cubicles were too small: 'Cubicles are very small and you end up climbing on top of the toilet to close the door'. In addition, one participant recommended providing shelves in toilet compartments so students could leave their books and bags on them instead of placing them on the floor. Another with visual impairment also commented that visual contrast between surface finishes and providing non-reflective surfaces as well as adequate lighting were important factors in facilitating access to people with visual impairments. 'I think having visual contrast in toilets is good practice. Having non-reflective surfaces is really good, to know where the wall starts and where it finishes'.

Wheelchair-Accessible Toilets

Participants with mobility difficulties and wheelchair users noted that although accessible toilets provided adequate space for manoeuvring, there were still a few issues that needed to be addressed: (1) heavy doors, (2) handles out of reach, (3) inadequate placement of grab rails, (4) high or low WC seats and washbasins, (5) knob taps that were difficult to operate and (6) soap dispensers out of reach.

One participant with mobility impairment commented on the relationship between the WC seat and washbasin provided in accessible toilets:

> The relationship of washbasin to WC pedestal is bad, there is a conflicting requirement, you want it not so far so somebody sitting on the toilet can reach the washbasin, but you don't want interference with somebody standing up, and now very often washbasins are used for standing and that will lead people to use washbasins as a hand hold and washbasins can come off.

In addition, taps caused confusion for many participants, especially those with visual impairments. Although all of them favoured the use of lever taps, the inconsistency of tap design, namely sensor, lever and separate taps, created confusion. One participant with visual impairment tried all the possible ways to operate the taps because there were so many different types of tap across the university. Another participant with visual impairment noted that when separate taps with no colour identification for hot and cold were provided, it was difficult to tell which was which:

> When taps are not labelled in red and blue, they are not very well placed and if you are not familiar with a certain toilet, you do don't know which is for which, because they either don't have a colour or it is not placed well.

Although accessible toilets were usually provided in separate compartments, some were incorporated inside the main toilet block, such as the ones provided at the Templeman Library Level 1 and Keynes College. One participant with visual impairment favoured provision of an accessible toilet in a separate compartment, rather than incorporating it inside the main block of toilets.

> I did not use an accessible toilet before. There is one in Keynes, which was incorporated within an actual toilet block, which I find quite difficult with my mother who is a wheelchair user. She had to open the main door then another toilet door, which was absolutely impractical. Accessible toilets should always be separate to avoid using to heavy doors.

Emergency Egress

Eight out of the ten participants interviewed noted that they had never submitted a personal emergency evacuation plan (PEEP), which is part of the fire-risk assessment for any building/premises for the evacuation of individuals with disabilities, and many of them knew nothing about it.

In addition, when they were asked whether they knew where the emergency exit routes were, five participants said that they were not familiar with these routes and many had never noticed them. Two participants mentioned that they had become aware of them because of their occupation, and three noted that they were familiar with some of the emergency routes in the buildings they used. In addition, seven participants pointed out that emergency exit signs were not clearly signposted and were hard to see, although three noted that the emergency signs were clear. Another participant stated that although emergency signs were clear, the number of signs allocated for each building was not enough. One participant also suggested that the university should improve the general access by providing level exit routes and evacuation lifts, and should improve signage.

> Emergency exit plans are impractical. If you are a student in the university, how will you know where you will be, for instance, at 3 o'clock the next day? It makes no sense to use it. A much better approach is to provide general access where the same level entrance routes are used as emergency exits and to provide evacuation lifts to accommodate the change in levels … Signage is not good and assembly points are not well signposted.

Finally, participants' responses revealed that 51% were aware of emergency exit routes, and pointed out that signage provided was clear (Fig. 5.24).

Management Procedures and Practices

Findings from the access audits and consultations with students and staff members with disabilities revealed that management practices and maintenance issues limited accessibility in some of the buildings. The main bad management practice that was noted was misuse of desks with low counters for storage purposes (Fig. 5.25).

One participant who was interviewed pointed out that although corridors in older buildings had the capacity to accommodate two wheelchairs and two pedestrians, furniture and fittings placed in corridors reduced the space.

> Fire extinguishers are most common then lockers and lift-up tables for people to write assignments … those are barriers because they can be visible for somebody who is upright, but one thing that you do as a wheelchair user, you look at the ground a lot, and I certainly run into these tables because I just did not notice them and suddenly you notice this hard edge going through you.

Online survey at the University of Kent
Question 17
Are you aware of the emergency egress/exit locations which operate in the building(s) you use?

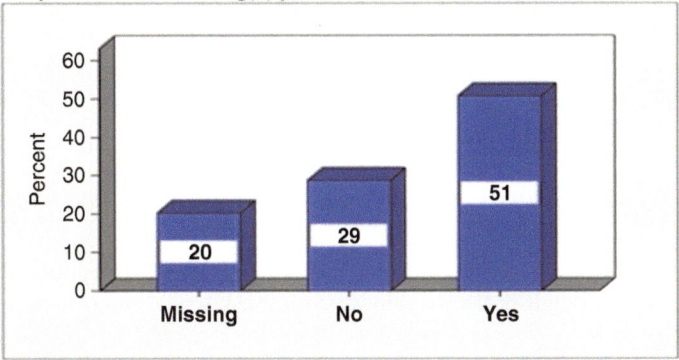

Question 18
Are the signs which mark emergency routes and exits clear enough?

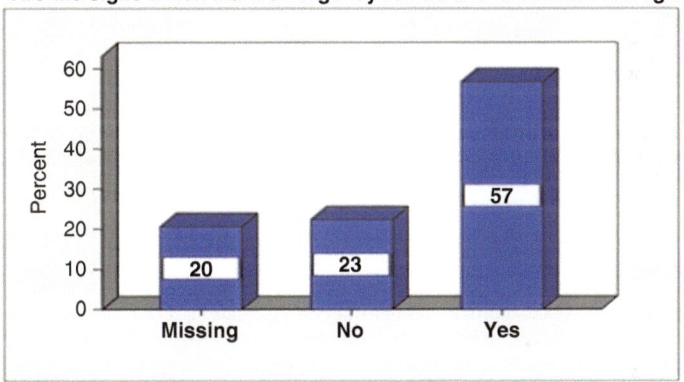

Fig. 5.24 Online survey results: emergency exit routes and signage

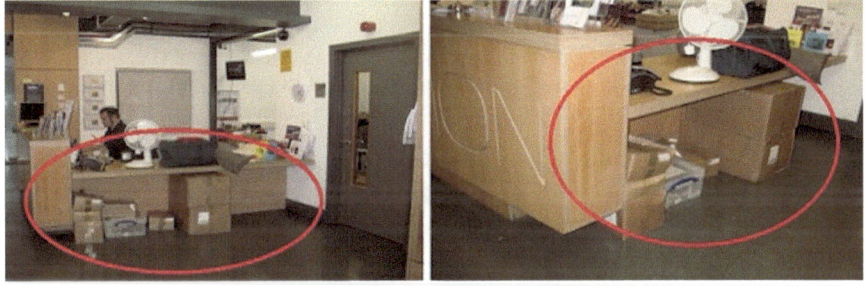

Fig. 5.25 Jarman Building low counter used for storage purposes

Staff Training and Disability Awareness

Although nine out of ten participants interviewed thought that staff members were helpful, friendly and polite, when dealing with them they pointed out that raising awareness among staff members about their specific needs was important in order to promote equality and inclusion. One participant, herself a staff member, pointed out that raising awareness among staff members about disability issues, such as the Access to Work Scheme, was necessary in order to improve services:

> In terms of their knowledge about accessibility, on the whole it is good, but I will say that, as a staff member, there is a lack of awareness about Access to Work, and I feel I have to really push everything to sort out everything by myself, and I have to take part in all of that. Staff members are supportive but do not have much knowledge.

Another participant thought that some staff members were exceptionally helpful because they had friends or family who were disabled, whilst others tried to accommodate the needs of students with disabilities by rescheduling classes to more accessible environments. One participant with mobility difficulties mentioned that although staff members tried to help and reschedule locations to offer accessible environments, the places allocated by the university were not totally welcoming and friendly.

> My supervisor has been excellent, in fact he has met me to discuss my PhD proposal; however, he met me in an open area, with a lot of chatting, and there was a band playing music, and I could not hear what he was saying. He was apologetic because he could not find a suitable accessible place. He was trying to accommodate my needs, but it was not ideal.

One participant with cognitive impairment pointed out that since his disability was hidden/unseen, either lecturers tended to be unaware of his problem or they dismissed it.

> Because I come across as normal, though I am on medication, lecturers don't sometimes understand my medical condition, which makes me get angry because I have acknowledged my disability to the Disability and Dyslexia Department, thinking that this data has been passed to lecturers so they can become aware of my condition .... this is my biggest complaint, if you have a database and you are not using it, then what is the point of having it in the first place?

A similar scenario was noted by a participant with visual impairment where lecturers were not aware of her specific need:

> Sometimes lecturers photocopy something and to save paper for environmental reasons, they do it small, and then you need to photocopy it larger. And I prefer it if they are aware that somebody in class needs it larger and that will save the student time.

One participant with visual impairment favoured services involving interaction with staff members rather than using 'high-tech' services, which could be quite confusing:

> In Rutherford College they installed a new catering system and it is based on using screens to order your sandwiches. And it is the most confusing system. It is touch screen and involves a very small TV screen to order your sandwiches on. Before, you used to go and

say, I want xx of sandwiches and you used to speak to a person, whereas it is now, as I call it, like an Argos system, it is like being in Argos, you are given a ticket with a number on and you have to wait for your number to be called, and it is on a tiny screen and when it gets busy you cannot see the screen very well.

When asked about facing any direct discrimination from students/staff members, participants noted that they did not meet it as such; however, they received some comments such as 'You are brave … Lucky to have such service … or something patronising. There were also looks that conveyed confusion'.

I don't face discrimination but I face confused looks, because when I look at somebody, I cannot look completely straight so I find people looking over my shoulder to see who I am looking at, and it can be quite embarrassing.

Effectiveness of Legislation

The online responses revealed that around 44% of respondents were not knowledgeable about legislation whilst 15% were aware of it. Although participants' responses to the online survey (Fig. 5.26) showed that around 16% were satisfied with the legislation, interviewed participants said that they understood the importance of the legislation, but the majority (eight out of ten) thought that the legislation was too broad and vague in its definition of the term 'disability'.

One participant thought that legislation was successful in terms of raising awareness about people's disability needs, but it did not have any power.

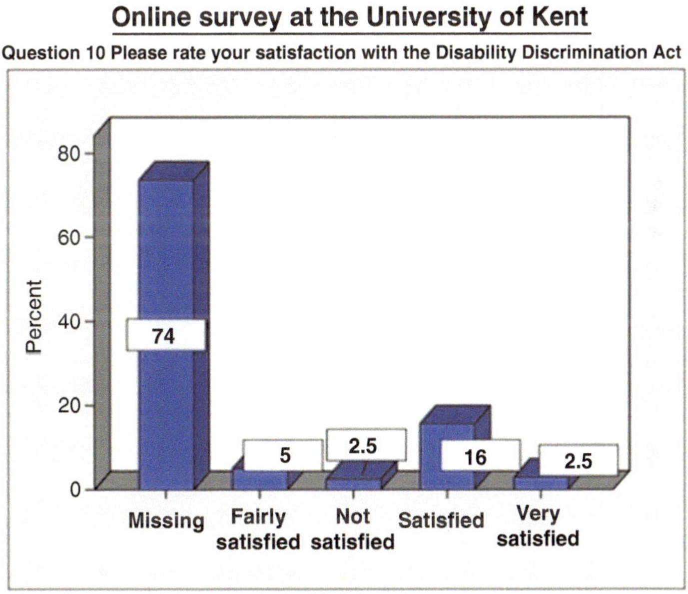

Online survey at the University of Kent

Question 10 Please rate your satisfaction with the Disability Discrimination Act

Fig. 5.26 Online survey results—legislation satisfaction

The killing piece is using the term 'reasonable', so who defines what is 'reasonable'? It is like using the term 'reasonable clothes'. And service providers can get out of anything. To make it more powerful, you need to define it clearly and make it obligatory.

Another participant commented: 'Legislation is good but things have to be changed, "reasonable adjustment" is broad and has to be narrowed. It does not mean anything and it does not explain what it means'. Similarly, one participant noted that the legislation was oriented towards limiting the extra costs of implementing change, and thus it used the term 'reasonable adjustment':

I think 'reasonable adjustment' is vague and what is reasonable adjustment for one person is not suitable for another, and I think they are all worried about costs, and again it does not have to be about cost, but about changing policy or procedure or training people. The DDA should be clear and more specific so there will be no room for different interpretations.

It was suggested by one participant that the legislation should include more sections in the Building Regulation standards that would incorporate solutions and adjustments to suit each disability. Three participants mentioned that people who were discriminated against could raise the issue in court to claim their rights; however, using the legislation and its broad definition of 'reasonable adjustment' to defend their case could be interpreted differently since there is no accurate definition of the term 'reasonable adjustment'.

The Equality Act reduces discrimination but it does not ensure it, because few people who are discriminated against at the workplace report it and many do not report it.

The Equality Act is not dealing with discrimination, not in definite terms, because there have been cases when people try to talk about x employees and suing and compensation and it is very difficult to prove it using the Equality Act to prove that discrimination has taken place and has been illegal.

The Equality Act does not have teeth to it. The only thing as I understand it, is that it allows a person to sue if they feel they have been discriminated against and they are waiting for the legislation to be reinterpreted in court and since the term 'reasonable adjustment' is not clearly defined, it can be interpreted differently ... in which makes it a murky area ... it is very subjective and it needs to be tightened up. There needs to be enforcement.

In addition, one participant pointed out that since the Equality Act was vague, those with mental health disabilities were not protected against discrimination.

I have heard many people saying that mental health discrimination is not covered clearly by the legislation. I know that people are pushing for a European Discrimination Act, and at the minute it is pushing stronger to include stronger parts to do with mental health. The legislation should include a section for mental health disability.

One participant suggested that the whole definition of the term 'disability' in the legislation failed to protect those people with a disability from discrimination and proposed that it should be amended.

Integration and Consultation with Individuals with Disabilities

The findings from personal interviews and online survey questionnaires conducted with individuals with disabilities suggested that integrating and consulting them in the course of design process and on site would improve accessibility. One participant recommended involving disability experts to a greater extent when implementing change: 'More disability expert representation, which means getting them more involved'. On the other hand, the same participant added that consultation with individuals with disabilities was not a common practice at the University of Kent, and although many people with disabilities tried to provide input and suggestions, such suggestions were not taken into consideration when refurbishing buildings or constructing new ones. 'I presented comments about the Woolf Building and none of my points were taken on board, and now they are facing problems and people are complaining about it'.

Another participant pointed out that it was important to present a reasoned argument to explain why things have to be done for all potential users.

> Many people with disabilities are not allowed to reality check, and don't know what is good and what is not. Effective feedback is important. Think about other people and other class members, etc., think about groups that are likely to be excluded and cater for their needs.

One participant commented when asked about how the university could create a friendly and welcoming environment that would suit all its students and staff members: 'If the University consults people and invites people and talks to them about their specific needs and talks to them about the design, then you are on your way of achieving that'.

Another participant pointed out that the University of Kent could incorporate accessible design by integrating the needs of all its users.

> It has to be more integrated. I think that is my biggest issue. A person with disability can get into the building, but if you are integrated with people, you would like to access the same door and all go in the same way. You don't have to be treated differently or access from a different entrance or use a special lift. Integration and inclusion is the key to incorporating an accessible built environment that all users can benefit from.

Inclusive Approach: Design, Policies and Practices

All interviewed participants commented that the University of Kent provided a friendly and welcoming environment for its students. As shown in Fig. 5.27, 80% of participants believed that the Canterbury campus provided the same level of accessibility for all its users. However, interviewed participants suggested that the university could promote inclusion by developing an inclusive environment that provided the following: (1) clear signage for all, (2) same access routes for all, (3) accessible toilet compartments that everybody could use, (4) consistent steps with tactile warnings and visual contrast, (5) automatic entrance doors with high-visibility features, (6) wider doors and (7) accessible parking.

One participant noted that integration and inclusion would be the key elements in providing an inclusive environment where: 'One would like to access the same door and all go in the same way. I think accessibility should be for everyone'.

Online survey at the University of Kent

Question 11

Do you think that you have the same level of access to Canterbury campus as anybody else?

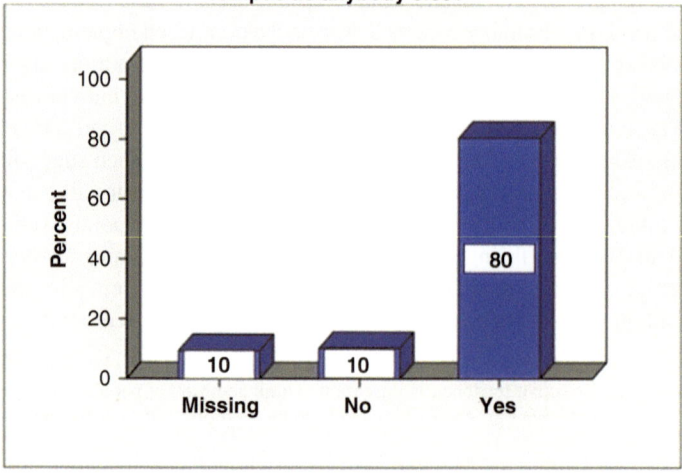

Fig. 5.27 Online survey results: evaluation of level of accessibility at the Canterbury campus

Another participant added, 'Providing clear signs so everyone will benefit because everyone is new and everyone has to find the place'. A further key factor that could promote inclusion would be to emphasise other disability needs, and not just mobility needs, according to one participant:

> I do think there is more emphasis on wheelchair use; there should be more attention to visual impairment or hearing impairment needs. They don't have other aspects mentioned, there should be more awareness about other disabilities and they should provide services that cater for their needs.

In addition, although participants favoured the inclusive design approach, many noted that effective feedback and consultation with individuals with disabilities were important practices that should be followed when adopting the inclusive approach.

> Inclusive design is like inclusive politics. One of the problems about achieving an inclusive environment at the universities is that not everybody is consulted. Many people with disabilities are not allowed to reality check … feedback has to be taken into consideration.

> Things should be made inclusive, and the university should talk to disabled people to incorporate their needs in their designs, rather than guessing what they might need without talking with them … so they should have consultancy widely with disabled people with different disabilities.

As shown in Fig. 5.28, 64% of survey participants favoured the inclusive design approach, whilst 11% preferred accessible design. Regarding design features, 62% considered that provision should be incorporated seamlessly within buildings,

Online survey at the University of Kent

Question 19 & 20

19- Which two design approaches below you recommend in the design process?

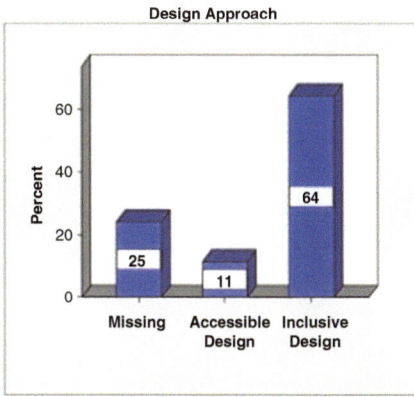
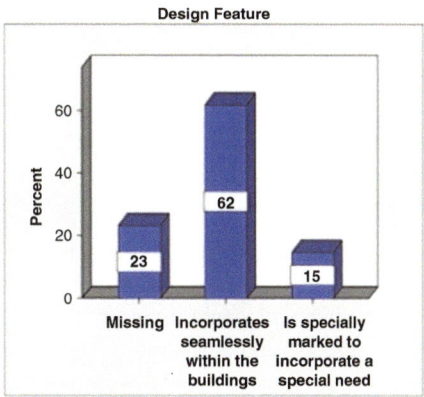

Fig. 5.28 Online survey results: favoured design approach

whilst around 15% thought that special provision should be made to cater for specific needs. These data revealed that the 11% who suggested accessible design fell into the 15% who favoured designs that were specially marked out for their specific needs.

5.5.3   Stakeholder Consultations

5.5.3.1   Overview

The purpose of the stakeholder consultation was to obtain feedback from education providers and architects, who worked on designing and constructing new buildings at the Canterbury campus, with respect to the effectiveness of the legislation/design standards and regulations for the provision of accessible environments and promotion of inclusion.

5.5.3.2   Results

This section answers the following research questions:

1. To what extent have education providers managed to make reasonable adjustments or eliminate architectural barriers?
2. To what extent are architects aware of the needs of potential users and people with disabilities when designing university buildings?

3. To what extent are architects aware of the inclusive design approach?
4. To what extent do architects abide by the design rules and regulations in their designs to cater for all users?
5. Has the guidance influenced architects' designs of new buildings or limited their creativity?

The stakeholder results are thematically presented with respect to the common themes that emerged after interviewing education providers representing four sectors: (1) Disability and Dyslexia Student Services (DDSS), (2) Equality and Diversity Department, (3) Estates Department and (4) Health and Safety Department. The views of two architectural practices were also obtained, and they shared their experiences of designing new buildings at universities. The themes were:

- Compliance with legislative duties
- Management procedures and practices
- Training, and disability awareness
- Integration and consultation with people with disabilities
- Inclusive approach in terms of policies, practices and procedures

Compliance with Legislative Duties

All interviewed stakeholders agreed that the legislation placed a duty on them to comply with the specific relevant disability regulations. Many believed that although a number of changes had been made to improve accessibility on university campuses, there were still issues that prevented them from catering for all users.

Education Providers

Many education providers adopted the Equality Act's definition of disability, but others noted that their understanding of the term disability was slightly different. One education provider commented, 'My view of disability is that it is either physical or cognitive disability and is identified by some form of medical diagnosis'.
Another education provider added:

> I have a slight disagreement with the DDA or the Equality Act definition of disability or the definition of best practice … I include people with sports injuries, broken legs, those recovering from surgery or who are heavily pregnant because they all require the same attention as individuals with disabilities in terms of emergency escape planning.

Accordingly, students with disabilities enrolling at the University of Kent needed to show DDSS medical evidence from their doctor or consultant, or if they had a learning difficulty or mental health condition, they had to provide a report from an educational psychologist. In addition, the DDSS offered a computer-screening dyslexia test for students who might be experiencing learning difficulties. When the test showed that it was probable that they were dyslexic, they were referred to diagnostic psychologists.

When asked whether they complied with the duties set by the code of practice to remove physical barriers for users with disabilities, education providers noted that improvements and adjustments to buildings had been made after carrying out access audits on buildings. Accordingly many lifts and automatic doors/power-access controls had been installed, and many accessible toilet compartments had been provided. However, a large number of education providers believed that there were many more access improvements required to cater for other disabilities, such as cognitive and mental difficulties. One stated, 'I don't know about how we manage learning difficulties and mental disabilities because most of that, I believe, is related to 'assessing the curriculum' and I believe that is dealt by a different support unit'.

Another education provider believed that the lack of descriptive regulations defining the term 'reasonable adjustment' was preventing stakeholders, architects and planners from removing potential barriers.

> I think broadly speaking, yes, we have managed to eliminate some barriers … The term 'reasonable' is a very difficult area and it is a bit of a movable phrase. It is like the notion of practical need and health and safety: it has some bearing on the financial ability of the organisation and cost benefit analysis, and that is a difficult bullet to bite in terms of a disabled person or group of people who need certain adaptations and suffer from discrimination or isolation if their needs were not met.

Architects

Personal interviews were conducted with two senior architects who worked for two well-established architectural practices in the UK. Architect X worked in a London-based practice formally established in 1988 and architect Y worked in a practice in London specialising in arts, education and public buildings.

Architect X viewed the relationship between architecture and disability as the necessity to study space, shape and time, and their interaction with people's social and cultural backgrounds. Whilst architect Y perceived an individual with disability to be someone who was less able to use standard facilities because his/her needs were not catered for, architect X pointed out that architects designing to take account of the needs of individuals with disabilities needed to address these requirements in relation to people's different lifestyles.

Architect X pointed out that disability was a mismatch between what was provided in society and what people experienced. When asked whether they had managed to cater for the needs of all users, including individuals with disabilities, both architects responded that they tried to accommodate these needs; however, it was hard to take account of all needs unless one discussed them with individuals with disabilities. Architect Y said:

> I think there are particular issues that people with certain disabilities face, in, sort of, one person is weak on one side, another on another side. Other people might have difficulty in staying upright, so there are many kinds of subtle details that make it difficult to do something without speaking to people to solve every potential problem.

Architect X noted that it was hard to be aware of every single individual need: 'It is very hard to say that our buildings cater for the needs of all people'. Architect X acknowledged that 'The wide range of disability makes it is very difficult to accommodate all these needs' suggesting that 'Accessibility should be better described, and I think that one has to try and find a kind of immediate or a point where you are trying to accommodate everyone so that the design becomes inclusive as much as possible'.

Management Procedures and Practices

This section presents the management procedures and practices of education providers and architects in offering their services to people with disabilities.

Disability and Dyslexia Student Services Department DDSS

A representative of the DDSS stated that arrangements for determining admissions for undergraduate students to join the University of Kent started with the UCAS application, which included a section about disability and specific needs. The Admissions Department would then feedback that information to the DDSS, which in turn would send out a registration form for students with disabilities in order to collect supporting documents or information, which would then be saved in their database. Once students with disabilities were granted an unconditional offer, the DDSS would write to students to seek further information and to make sure that they had applied for a student disability allowance from the Financial Department of the University. The student disability allowance in its turn would fund equipment that the student might need. In addition it would fund salaries for those providing human support, such as a dyslexic tutor or mental health mentor, and educational support assistants. The DDSS office deals with students who sometimes:

> Come and share their architectural barriers and then we contact the Estates Department to remove these barriers. We have got the Disabled Go access guide, although it is not something that meets the requirements of the legislation, but it identifies for us where there are access issues.

With respect to some students with mobility difficulties who did not have their own facilities, the DDSS would provide mobility scooters for loan on campus. A representative of DDSS pointed out that accessibility barriers were dealt with by the Estates Department which prioritised actions and allocated funding resources to implement necessary adjustments. However, there was also a little financial support that came to the University from higher education disability funding. The DDSS conducted an annual survey of all registered students within the department to get their feedback on services and support provided.

> We receive their comments about university experiences where they mention good things and difficult things, and then we feed it back to our own department and also to other departments, such as issues related to parking and library. Surveys have influenced in planning ahead.

When asked about the recruitment polices for encouraging students to apply to the university, the representative of the DDSS pointed out that the university marketing prospectuses, Disabled Go website information and open days were used to attract applications from disabled students. In addition, some positive experiences of students with disabilities had encouraged others to apply to the Canterbury campus: 'We have increasingly students coming to us with Asperger syndrome, for example, the feedback from existing students and parents have been positive and that encouraged a greater number of applicants in that area'.

Equality and Diversity Department

The Equality and Diversity Department had duties to provide equal opportunities for staff members, and dealt with staff issues more than with student issues. Recruitment was an important process for the Department, which provided an online system known as 'Eye Grasp' through which potential staff could find advertisements for vacant posts and submit applications.

> As part of that process we look at disabilities as well. For example, we are part of the scheme called the 'two ticks system'. It is a cognitive initiative, and basically it means that if a disabled member of staff applies for a job and they meet all the essential criteria of the job, then they are guaranteed an interview, so it is a positive action initiative and is part of the recruitment process.

However, the representative of the Equality and Diversity Department pointed out that although the job description contained a section about equality and diversity and welcomed staff from all areas of the community, the Department was not doing enough to attract applicants with disabilities: 'We do not have a specific procedure or media to advertise in any specific disability media'. When asked whether the Department evaluated its services to staff with disabilities, a representative stated that they never conducted surveys to obtain feedback from individuals with disabilities; however, the Department conducted an equality impact assessment of policies and procedures and assessed them on an equality assessment basis and in terms of ways in which it affected people with disabilities or any particular group.

> We have started that last year, probably September, and in a way it is an audit. We are doing human resources policies like probation policies etc. and we are looking at these things. And also we are looking at the lack of data and we are working on these things.

Fire Safety and Environment Department

The Fire Safety and Environment Department was in charge of the emergency provision for people with disabilities in terms of organisational and managerial issues. Such provision included dedicated signage, door entry systems, refuges and providing colour contrast in staircases. To comply with its duties to provide reasonable adjustments, the Department provided a fire emergency plan for each building with special arrangements for evacuating users with disabilities, such as providing flash beacons, vibrating pagers and pillows for those with hearing impairments, steps with colour contrasts for visually impaired people, and evacuation chairs, as well as one evacuation lift in the Jarman Building for people with mobility difficulties. The Department had contacted disabled students residing in campus requesting them to answer three questions:

> Are you able to proceed and respond to a alarm system?

> Are you able to evacuate in five minutes, and this varies depending on size of building? And, thirdly, in terms of disability etiquette, this is very slightly tricky, in a self-evacuation might you impede other people evacuating?

If a student was likely to experience difficulty in evacuating independently and safely, an appointment would be arranged to discuss their needs.

> Usually drawing on generic possibility or whatever additional needs that they might have, any consultation with them are a joint process, it is not me saying "You going to do this". We draw up an individual personal emergency evacuation plan that we call 'PEEP'.

In addition, the fire safety representative pointed out that although many might feel that arranging a self-evacuation plan was exclusive and would promote segregation, it was a necessity to protect all in case of emergency.

> I have to give an advice on the main flow and evacuation exits for everybody including possible danger to person with a disability. It might seem a little unfair or segregation but, it is just really applying necessity.

The Department believed that Personal Emergency Evacuation Plans (PEEPS) which were used in halls of residence were so useful that it might consider offering them to other buildings across the University.

> 'PEEPS' are very useful, as I say, candidly, we have done quite a few but we need to do a lot more, not only residence buildings, but more buildings, and we start with all regular users of all buildings and that is what we aim for in the future.

Estates Department

The Estates Department was in charge of assessing buildings, setting priorities when it came to implementing adjustments, and also funding adjustment and building costs. A representative of the Department mentioned that most projects were centrally funded by the government; in addition some funding was granted by the DDA in the first 2 years of legislation to encourage the removal of architectural barriers. When asked whether the funding was sufficient to remove such barriers, the Estates Department representative stated that the university was trying its best to remove them, though much work had still to be done once funding was in place.

> Currently from access audit we are significantly away from having everything done on the floor. A lift can cost us between £60,000 and £100,000 to pay for the structure that goes around it. If you have a quarter of million pound budget, it does not go that far.

He also noted that most students' residential rooms were fully adapted to cater for the needs of individuals with disabilities; however, adapted rooms were always placed near each other, which undermined inclusion.

Unfortunately, due to designers, we tend to group students with disabilities on a corridor where they have got much bigger bathrooms and doors are wider, power access is available from coming in all the way to their bedrooms.

Architects

Architects X and Y acknowledged their duty to accommodate the needs of individuals with disabilities. The two interviewed architects stated that access statements and Part M were their starting point when designing a building for individuals with disabilities.

'We look at Part M, and that is from really early days, just to make sure that we are building that into design along with other kinds of requirements'.

When asked about particular problems they faced when designing for users with disabilities, both architects confirmed that they had never faced any problems; however, they mentioned some other features that were problematic for users with disabilities. Architect Y stated:

Lift control: people with particular disabilities might not have a good catch control. Heavy doors are a real issue, particularly with the air-tightness of buildings these days, and that is kind of difficult to achieve, but there are not any other major issues. Some of the colour-contrast things, to get enough contrast that people can see switches, distinguishing the walls and floors, but without the building being kind of multicoloured.

Architect X said that designing for individuals with disabilities was not a challenge; the challenge was successful communication between architect and client.

It is about communication between architect and client. It is about things that affect the person's prejudice and involve main confrontations ... It will be fair to say that people are frightened or there is some misunderstanding or you, as an architect, just have to confront your own prejudice in order to participate in that whole process.

When asked whether their designs complied with British standards, the two architects stated that they audited their plans to make sure that all designs were in accordance with standard measurements. Architect X mentioned that involving access consultants and individuals with disabilities in some projects was another way of checking compliance with British standards, stating: 'Access consultants and individuals with disabilities would be involved in reviewing the project in each stage, and sometimes there is a client who appoints them, and sometimes they are part of our design team'.

On the other hand, the two architects believed that although their designs complied with British standards, there were some features which created real barriers: lift designs proved to be the main barrier for most users with disabilities. Architect Y noted: 'Obviously if there is something that has to be amended, maybe it is the lift design and there are issues with it, and it makes sense to do it'.

Architect X pointed out that since the university had its own standards and checklists that it used to cater for particular needs, it sometimes required all users to be able to evacuate the building, and so adjustments had to be made when a potential user with another disability or need had to use the building. This in turn added more costs to the construction budget. Architect X added: 'I think that there are ways in which it is not ideal because you are then, kind of, facing this kind of altering to accommodate something, which is not nice'.

Training and Disability Awareness

Interviewed stakeholders reported that they provided their employees with some form of disability awareness training; however, such training differed from one department to another, depending on the services provided.

DDSS Training

DDSS provided its staff members with a staff development programme run by human resources. The course was known as 'Equality and Diversity', and within it there was a disability awareness module that provided training in initial disability awareness. In addition, the DDSS provided a training course known as 'Disability Etiquette' in which staff were trained to use language correctly and learn about many other related issues. The training course asked staff to work in pairs, and they looked at issues that affected people with visual impairment, hearing impairment and speech impairment and discussed how to help people to overcome those barriers by working on a one-to-one basis.

Moreover the DDSS invited a few staff members to deliver training courses in which they talked about their experiences of disability and working in a university. 'It is always very successful, it is a kind of empathising with people with a range of difficulties and helping them, and that is part of the training'. There were also training courses about legislation offered by the University for senior managers, heads of schools, heads of sections and executive groups, which they in turn passed on to their own staff.

On the other hand, a DDSS representative highlighted the fact that there was a lack of training courses that addressed recruitment activities, which aimed to reach individuals with disabilities, and stated that there was a need to provide more specific training courses dealing with mental health difficulties.

Equality and Diversity Training

The Equality and Diversity Department offered a disability awareness course once a term. The course also provided brief information about legislation and discussed different types of disability and ways to provide support to cater for the needs of disabled people.

The Equality and Diversity Department also provided an annual recruitment training course for staff on recruitment panels, and this included a section dealing with disability.

Fire Safety and Environment Training

The Fire Safety and Environment Department delivered training courses on health and safety issues concerning disability. One new course provided training for staff members in how to prepare individual personal evacuation plans, and also contained disability and equality awareness information. The Department also took part in training courses covering aspects of fire emergency response training for security officers and fire marshals. In addition, deputy directors and health and safety advisors attended the equality and diversity training course offered by Human Resources. However, a representative of the Fire Safety Department acknowledged the need to provide separate training courses about disability. 'It is interesting to have a separate feature on disability training and provision in terms of evacuation, which, you could argue, is not inclusive but it is just a necessity'.

Estates Department Training

A representative of the Estates Department stated that he had never had any training related to disability awareness, legislation or recruitment that targeted individuals with disabilities; however, he was quite familiar with disability since he had personal experience of a relative who had a disability.

Architect Training

Whilst architect Y had never had any disability training in his architectural practice, architect X had attended a training programme within the architectural practice and was informed about changes to legislation and general standards. He stated:

> We also work with access consultants who teach us a lot and that, I guess, is another form of training. And they are often appointed by our clients as well to ensure that the building that has been designed meets their needs.

Integration and Consultation with Individuals with Disabilities

Education Providers

The education providers said that they integrated people with disabilities in their training courses, but students and staff members with disabilities were never consulted once physical adjustments had been made to certain buildings. One interviewed education provider acknowledged this, saying, 'I think we could do more, we could consult students and staff members who use the services, and that is something we don't do at the moment'. Another education provider wondered: 'Why do barriers straight away become acknowledged immediately after a new building door is opened? It could be because they don't consult individuals with disabilities'.

Architects

Similarly, architects X and Y pointed out that seeking advice from individuals with disabilities was important in order to accommodate their needs. Architect X said:

> Depending on the type of building that we are designing, we would speak to people with disabilities as we are developing the design, particularly around aspects of the design that are particularly problematic for a specific user or that it might cause particular difficulties for people with disabilities.

Architect Y commented that consulting with a client with a disability during the design process was useful in terms of creating a space that would cater for the client's specific needs. In a similar scenario architect X stressed the idea that consulting clients, no matter what their abilities were, and engaging them from the beginning were essential to enabling both architects and clients to achieve accessibility.

Inclusive Policies, Practices and Procedures

Education Providers

Education providers complained that accessibility was always biased towards wheelchair users, whilst other disability needs were neglected.

> There is a tendency to think about wheelchair access, but there are other situations as well, like people with visual impairment and the issue of colour contrast, and there are other things that are forgotten. For example, signage, physical access is not always thought, and they cannot even get that right.

Another education provider added:

> Although we have a disability consultant, what tends to happen is that most of normal types of people with disabilities are wheelchair users, so we really focus on them because they need more help. We think that someone with a hearing disability can function within normality, and they don't need anything particularly special, except induction loops.

Accordingly, all interviewed education providers considered that inclusive policies, practices and training helped to improve the quality of services at the university. One provider referred to 'combination of awareness, willingness, understanding and funding that is all part of that really improving the quality of environment'. In addition, a representative of Equality and Diversity suggested that there should be more cross-communication between departments, saying that there should be 'communication with Estates and our department and DDSS, we could use more specialists in disability and it will help a lot, or disability advisors'.

Architects

Both interviewed architects acknowledged the importance of incorporating inclusive designs that would take into consideration the needs of all potential users. Architect X defined inclusive design as follows: 'It is a design that is easy for everyone to use and does not kind require special facilities for people with disabilities where that it is possible'. On the other hand, architect Y pointed out that the big conflict in adopting an inclusive approach arose when working on existing buildings, which could be quite challenging.

> I guess the most difficult situations are when you are working with an existing building, especially when that building is laid out on multiple levels, trying to get access to all or at least to a very large majority of the building …

Another important factor that architect Y valued was cooperation between access consultants and architects, as this assisted in overcoming barriers and promoting inclusion.

> An access consultant is really an important member of the consulting team and we need to involve her in the tough issues of moving people to the appropriate level without ending up with 20 lifts, but providing access to all levels of the building.

References

Finkelstein, V. (2002). The social model of disability repossessed. *Coalition, February, 10–16*

Sawyer, A., & Bright, K. (2007). *The access manual: Auditing and managing inclusive built environments*. Oxford: Blackwell Publishing Inc..

Swain, J., & French, S. (2000). Towards an affirmation model of disability. *Disability & Society, 15*(4), 569–582.

Chapter 6
The Impact of Legislation on University Buildings: Inclusive Design Proposals

6.1 Introduction

There has been growing awareness of equality and disability rights because of legislation that protects people from discrimination, such as the UK Disability Discrimination Act 2005, which was replaced by the Equality Act 2010. This awareness has caused a shift in attitudes in favour of inclusive design. The legislation gives rights to individuals with disabilities, and these include equal opportunities in the areas of education, employment, access to goods, facilities and services, and buying or renting land or property. Inclusive design means designing products, services and environments that as many people as possible can use, regardless of age or ability. It is also known as 'universal design' or 'design for all'. Inclusive design is a new attitude or approach to design in general and not a new style of design (Burton and Mitchell 2007, p. 5–7).

The importance of inclusive design or access for all has become a mainstream concern. Recent demographic studies have shown that by 2020 half of the UK adult population is expected to be aged over 50, which means that the built environment has to cater for this group so that it is not discriminated against. The adoption of the social model of disability aims to promote inclusive environments that meet the needs of all people, irrespective of their different abilities and ages. An important aspect of creating an inclusive environment also lies in the attitudes and values of architects, developers and service providers and their underlying motivations in acknowledging the needs of their clients.

This chapter presents three case studies of British universities founded in the 1960s to investigate whether Approved Document Part M has impacted the building designs to achieve inclusive design. Six buildings, each from a different decade, namely the 1960s, 1970s, 1980s, 1990s, 2000s and 2010s, located at the Universities of Essex, Bath and Kent, were compared to assess the level of compliance with Part M, and how each building design has responded to a wide spectrum of users' needs, including those of individuals with disabilities. The aim of this examination is to

© Springer Nature Switzerland AG 2020

I. Shuayb, *Inclusive University Built Environments*,
https://doi.org/10.1007/978-3-030-35861-7_6

find out how each university tackled the accessibility barriers to respond to potential users' needs.

Moreover, the chapter provides strategies for designing inclusive environments with respect to buildings at the University of Kent. Two different case studies both centring on the inclusive design approach offer the reader new insights into the way people interact with the built environment. The two case study buildings were selected to take account of building type, period of original construction, usage, and historical and organisational type.

This chapter presents two main categories: existing buildings and new buildings. Existing buildings include the first ones to be erected and those constructed before the study was conducted. New buildings are those buildings that have been built when the study was conducted.

It is important to note that the possibility of achieving inclusive design in new buildings is more likely to be achieved if the designer has the opportunity to apply the inclusive design criteria from the preplanning through to the construction and occupancy phases. However, the challenges of enhancing accessibility and achieving inclusive design in existing buildings are greater as the designer is constrained by existing features, which have to be altered to make them more accessible for users with different abilities.

Historic buildings are also categorised as existing buildings; however, the constraints of enhancing accessibility and achieving inclusiveness in such buildings are even more challenging than in the case of most other existing buildings as the designer has to preserve the exterior facade of historic buildings to maintain their special architectural and historic character. England recognises the need to conserve the specific characteristics of historic buildings and lists them according to their period of construction. Historic buildings, as defined in Approved Document M, include listed buildings; buildings situated in conservation areas; buildings which are of architectural and historical interest and which are referred to as a material consideration in a local authority's development plan; buildings of historic architectural interest within national parks and world heritage sites; and vernacular buildings of traditional form and construction (Approved Document M, 2004, p. 14).

Acknowledging all these constraints, these case studies reflect how the inclusive design approach can be applied to different existing and historic building types to achieve inclusiveness. They also demonstrate how inclusive design incorporates design that promotes health and safety, and social participation, without affecting the existing building fabrics.

The Templeman Library case study describes and evaluates the level of accessibility of an existing building founded in the early 1960s at the University of Kent. The Library is used by University of Kent students and staff members, in addition to visitors who use the building to study, read, work and socialise. The Templeman Library was going through a process of extension and potential reconfiguration of the existing space when the study was conducted, and this helped the author to propose an inclusive design solution that would tackle the needs of all potential users. The Templeman Library case study is used as an example to show how an existing 1960s' building going through a process of extension offers an opportunity to achieve inclusive design.

Eliot College, erected in the early 1960s, was one of the first university buildings to be constructed at the University of Kent. The College, which is used by students and staff members as work, study, accommodation and social entertainment space, illustrates how the issues concerning orientation and navigation can be resolved by applying inclusive design principles without going through major structural alterations or extensions.

6.2 Impact of Legislation on University Buildings

This section examines the success of legislation and Approved Document M in removing architectural barriers for individuals with disabilities at the buildings of three British universities founded in the early 1960s. Tables 6.1 and 6.2 illustrate the impact of legislation and Part M, showing how each building removed the accessibility barriers for its users, including individuals with disabilities.

The review presented in Tables 6.1 and 6.2 aims to reveal the relationship between the legislation and the changing design of university buildings over time. Findings from the literature review and Tables 6.1 and 6.2 show that there is a practical dynamic between the legislation and the nature of university design and the impact of the legislation on meeting accessibility requirements.

The findings displayed in Tables 6.1 and 6.2 and the access audits presented in Chap. 5 reveal that although legislation and accessibility standards have improved access to university buildings, they have failed to achieve inclusive design. It is clear that the legislation during the 1960s and 1970s had no impact on improving accessibility at universities. Its impact became effective in terms of eliminating physical barriers, but only for wheelchair users, in the 1980s and 1990s when ramps and passenger lifts were provided to enhance accessibility for people with mobility impairments. An interesting finding, as shown in Table 6.1, highlights the fact that the amended legislation in 2005 and 2010 had an influence on university buildings constructed since 2000 in terms of their accessibility to users with mobility and visual impairments, but it has failed to achieve inclusive design that anticipates the needs of a wide range of users, including people with hearing or other cognitive impairments, children and elderly users.

The findings highlight the fact that there are many constraints in the practice of design that undermine the ability to achieve the provision of limited inclusive environments in university buildings. A key constraint is that although the current legislation requires universities to improve accessibility in their built environment, legislation before the DDA 1995 did not demand a review of design proposals at any government level. Although access improvements were implemented at the Universities of Bath, Essex and Kent during the 1970s, the changes that were made were not implemented sufficiently well to comply with the previous Approved Documents, nor did they comply with the current Approved Document M. This limitation is particularly found in older buildings at the three universities.

Table 6.1 Impact of legislation on university built environment

Legislation	Buildings	Impact of legislation on access improvements
1960s British Standard Code of Practice on access for the disabled to buildings CP96 in 1967 (public unisex accessible toilet with 1370 × 1750 mm)	Eliot College, Kent	No impact, building was inaccessible to individuals with disabilities.
	South 4, Bath	No impact, building was inaccessible to individuals with disabilities.
	Mathematics Department, Essex	No impact, building was inaccessible to individuals with disabilities.
1970s Chronically Sick and Disabled Persons Act (1970) Section 4 Providing access to public buildings without introducing provisions	Templeman Library extension, Kent	Gradient entrance from back. Telephone placed at back entrance where students had to call porter to help them enter the library.
	Arts Barn Centre, Bath	Ramp.
	Student union bar, Essex	No impact, building was inaccessible to individuals with disabilities.
1980s Building Regulation 1985 Part T Only single-storey buildings had to be accessible to disabled people.	Registry, Kent	Ramp, unisex accessible toilet, lift.
	West 8, Bath	Ramp, unisex accessible toilet, lift.
	School of Law, Essex	Level access.
1990s Disability Discrimination Act (1995) Education was included under the DDA 1995. Put duties on education providers to eliminate physical barriers to comply with Building Regulation Approved Part M 1998	Templeman Library extension, Kent	Passenger lift.
	The Library, Bath	Level access, passenger lift.
	Square 1, Essex	Ramp, passenger lift.
2000s Disability Discrimination Act (2005) Recommendations on car parking, access routes, entrances and interiors, horizontal and vertical circulation, surface finishes and communication aids.	The Venue, Kent	Level access, passenger lift, staircase with nosings, accessible toilet.
	New Chemistry Building, Bath	Ramp, passenger lift, staircase with nosing-accessible toilet.
	Building 2001, Essex	Level access, ramp, staircase with nosings, passenger lift, and accessible toilet.

(continued)

Table 6.1 (continued)

Legislation	Buildings	Impact of legislation on access improvements
2010s Equality Act (2010) Eliminated physical features and provided auxiliary aids to disabled users.	Jarman Building, Kent	Level access, automatic doors, counter desk with low counter, passenger lift, staircase with nosing-accessible toilet.
	East Building, Bath	Level access, automatic doors, passenger lift, staircase with nosings, accessible toilet, baby changing facilities.
	Tony Rich Teaching Centre, Essex	Level access, automatic doors, staircase with nosings, passenger lift, accessible toilet.

Access statements were first introduced in the Planning and Compulsory Purchase Act 2004 and took effect from August 2006 (Sawyer and Bright 2007, p. 105) as an exemption from compliance with Part M, which caused some confusion at the time. In May 2006, the government introduced changes to the planning application process and required access statements to be included in planning applications effective from 10 August 2006. To ensure that reasonable provisions were made for access to all users, access statements became part of the planning application requirements for extensions, listed buildings and new constructed ones at both planning and Building Regulation application stages. Moreover, Approved Document M (2004) of the Building Regulations contained a section that required access statements and proposed plans to be submitted for approval or once a building notice is obtained (Building Control Guidance Note 2016).

Whilst an access statement is seen as a way of achieving an inclusive environment for building, by ensuring continuity throughout the process of planning and design and the management of buildings and spaces, a review of the access statement for the Jarman Building at the University of Kent revealed that many of the inclusive features had not been completely adopted. This may happen when there is a lack of follow-through after design implementation. Although access statements present the opportunity for architects and designers, developers, and managers of environments to show their willingness to provide accessibility provisions in the work they undertake, they must meet and continue to meet the various legislative obligations (Building Control Guidance Note 2016). Review of the access statement and access audit assessment of the Jarman Building at the University of Kent revealed that access statements were not fully implemented during the construction and execution phases. Sawyer and Bright (2007, pp. 35–36) highlight the importance of a participatory approach to preparing access statements in the preliminary stages of the development process to make sure that the design proposed and its implementation will respond to the users' needs. However, findings from interviews with individuals with disabilities highlighted the fact that consultations with end users, including individuals with disabilities, had never been carried out in the preparation of the Jarman Building access statement, nor did any take place after occupancy of the building.

Table 6.2 Impact of Approved Document M on university buildings

	Buildings	Impact of Part M										
		Accessible parking bays near building	Ramp	Level access	Automatic main entrance door	Staircase with nosings	Passenger lift 1400 × 1000 mm	Counter desk Low counter	Platform lift 1250 × 800 mm	Accessible toilet	Ambulant cubicle	Baby changing facilities
1960s	Eliot College, Kent	✓	✓			✓			✓	✓		✓
	South 4, Bath	✓	✓			✓			Does not have one	✓		
	Mathematics Dept, Essex			✓	✓	✓		Does not have one	Does not have one	✓		✓
1970s	Templeman Library extension, Kent	✓	✓			✓	✓			✓		
	Arts Barn Centre, Bath		✓									
	Student union bar, Essex		✓		✓	Staircase with no nosing	✓	✓	Does not have one	✓		
1980s	Registry, Kent	✓	✓		✓	✓		✓	Does not have one	✓	✓	✓
	West 8, Bath		✓			✓			Does not have one	✓		
	School of Law, Essex			✓	✓	✓		✓	Does not have one	✓		

| Buildings | Impact of Part M | | | | | | | | | | |
	Accessible parking bays near building	Ramp	Level access	Automatic main entrance door	Staircase with nosings	Passenger lift 1400 × 1000 mm	Counter desk Low counter	Platform lift 1250 × 800 mm	Accessible toilet	Ambulant cubicle	Baby changing facilities
1990s Templeman Library, extension, Kent	✓	✓		✓	✓		✓	✓	✓	✓	✓
The Library, Bath			✓	✓	✓			Does not have one	✓		
Square 1, Essex		✓		✓	✓			Does not have one	✓		✓
2000s The Venue, Kent	✓		✓		✓			Does not have one	✓		
New Chemistry Building, Bath	✓	✓		✓	✓		Does not have one	Does not have one	✓		
Building 2001, Essex			✓	✓	✓		Does not have one	Does not have one	✓	✓	

(continued)

Table 6.2 (continued)

	Buildings	Impact of Part M										
		Accessible parking bays near building	Ramp	Level access	Automatic main entrance door	Staircase with nosings	Passenger lift 1400 × 1000 mm	Counter desk Low counter	Platform lift 1250 × 800 mm	Accessible toilet	Ambulant cubicle	Baby changing facilities
2010s	Jarman Building, Kent			✓	✓	✓		✓	Does not have one	✓	✓	
	East Building, Bath			✓	✓	✓	✓	Does not have one	Does not have one	✓		✓
	Tony Rich Teaching Centre, Essex			✓	✓	✓	✓	Does not have one	Does not have one	✓	✓	

To achieve an inclusive built environment, a participatory approach and consultations with end users with different abilities and needs are vital once preparing an access statement and end users should be consulted at each stage of development, including the design and implementation phase, and anytime a building or space undergoes a construction change that may affect the accessibility level for its potential users (Building Control Guidance Note 2016). Moreover, to ensure that access statements are implemented throughout the planning, construction and management of buildings, a monitoring board or team under the building control should review and check that such implementation is being carried out. Following up on the implementation phase is vital in order to monitor and check that the inclusive design criteria have been adopted.

Another important constraint when it comes to achieving inclusiveness at university buildings is the poor implementation of accessibility audits and assessments. The ramp at Eliot College at the University of Kent and the internal ramp in the Mathematics Building at the University of Essex are two examples that show how design has failed to comply with accessibility regulations. Although these universities think that they have done their duty in making provision for accessibility to comply with the legislation, they have not checked carefully that these provisions fully comply with design regulations.

Universities aiming to achieve inclusive design and comply with the legislation should check that implementation of accessibility audit recommendations is accurately undertaken. Quality control of the executive team is vital in checking compliance. Access consultants and inclusive design specialist teams should follow up and check on the implementation compliance. Making accessibility a shared task is essential to the achievement of inclusive environments. Proper training of the executive team with respect to accessibility and inclusive design is important in the implementation phase. Achieving an inclusive environment should be the equal responsibility of access auditors and consultants, inclusive designers, architects and executive teams to ensure that inclusive design becomes part of the university's culture.

One main legislative constraint is that the design regulation Approved Document M only applies to existing buildings undergoing extension or material change of use, and to new buildings. It does not include external routes and environments through which users need to pass to reach buildings. The Arts Barn Centre at the University of Bath is an example of an existing 1970s' building that has not undergone an extension and thus has not incorporated accessible provisions. Whilst it is not required to comply with Approved Document M as it has not undergone an extension, the services provided within it should not discriminate against individuals with disabilities (Equality Act 2010). However, an access audit of the building revealed that it was inaccessible to wheelchair users as it did not provide level access, nor a passenger lift. The main entrance doors were heavy and could not be operated by people with limited dexterity. Moreover, its staircase did not cater for individuals with visual impairments since its steps did not have clearly marked nosings. Another accessibility limitation of the building was the absence of accessible toilets and ambulant cubicles as well as a lack of baby changing facilities.

Establishing inclusiveness should be a priority according to the legislation. Inclusive design criteria should be reflected in Building Regulations and applied to all building types, regardless of the age of the building and the condition of the construction. Universities aiming to achieve inclusive design should not only adopt accessibility standards to comply with their legal duties, but also have a proactive strategy that will reduce the probability of future accessibility barriers in order to ensure a satisfactory response to the wide spectrum of users' abilities and cultural differences.

From an examination of the access improvements at selected buildings, it is clear that inclusive design is still not recognised. Legislation and Building Regulations have impacted the accessibility level with regard to university buildings but have not achieved an inclusive built environment. Sections 6.3–6.4 present two selected case studies that demonstrate how inclusive design incorporates design that promotes health and safety and social participation without affecting the fabric of the existing buildings.

6.3 Templeman Library

6.3.1 Overview

The Templeman Library, named after the university's first Vice Chancellor, is located at the centre of the University of Kent's original Canterbury campus. The Library was designed by Lord William Holford, President of the RIBA in 1964, who designed the master plan of the university. In 1968, the Library's central and western blocks were built, whilst part of the east wing was completed in 1974, an extension to which was added in 1990 (Martin 1990). Figure 6.1 presents the construction phases of the Library.

In 1974, computers were installed at the Templeman Library, and it became the first library in a UK university to provide online facilities.

The Library contains over a million items, including books, journals, videos, DVDs and archive materials. It also provides services, facilities and resources to support students, staff and visitors in the Canterbury campus. Moreover, the Library stocks the British Cartoon Archive, established in 1975, and includes a national collection of newspaper cartoons with over 90,000 catalogued images. It is estimated that around 800,000 people visit the library per year, with approximately half a million loans *per annum*.

The Library opens on Monday to Friday from 8:00 am till 3:00 am and from 9:00 am till 3:00 am at weekends. Students can only access the Library after 9 pm by using their (Kent One) ID card, whilst visitors are not permitted to access the Library after that time. After 9.00 pm the Library is self-service. The Library does not hold books in Braille as a matter of course, nor does it provide books in large print, but it can provide books on tape or disc. Moreover, the Library offers a book-ordering service, which takes 1–14 days to supply an order. In addition the Library provides computers with accessible software, and auxiliary aids are available. Although the Library makes some computers with keyboards available for visually impaired users, it does not provide large-text keyboards or large rollerball mice for users with visual

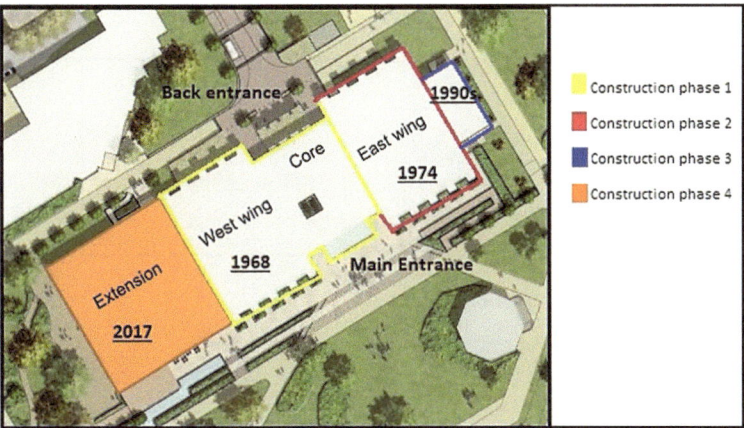

Fig. 6.1 Templeman Library construction phases

impairments. Equipment available for people with a visual impairment includes a text-recognition scanner, and coloured paper for printing and photocopying. As regards furniture, there are height-adjustable tables for computer users. Students can renew books remotely by phoning Lending Services or online via the university website. Moreover staff can photocopy from books, prepare course material, retrieve books from the shelves and organise library tours at 24-h notice.

Students and visitors can reach the Templeman Library either by using the bus services to Keynes and Darwin Colleges or by car, parking being provided at the rear entrance near the Gulbenkian Theatre. People with disabilities can also park their cars at the two designated accessible parking bays at the back of the Library (Fig. 6.2).

6.3.2   Templeman Library Accessibility Barriers

The Library has five floors (basement, Levels 1, 2, 3 and 4), each covering the central, east and west blocks. Findings from access audits, interviews with students and staff members with disabilities, and real-time observation of the flow of Library users showed that the Library was inaccessible to the same potential users, including people with disabilities, which resulted in exclusion and segregation. The first accessibility barrier noted was the absence of level access at the main entrance, with the reception counter being placed on Level 1, which could only be accessed by internal steps or a glazed platform lift (Fig. 6.3).

Although the platform lift provided access to Level 1 and the basement, it promoted an exclusive environment since it was specifically provided to serve people with mobility impairments and wheelchair users without acknowledging the diversity and differences in users' abilities and needs, which is one of the principles of

Templeman Library Car Parking

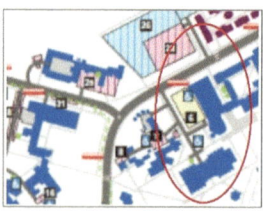

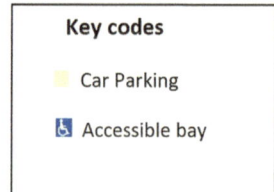

Key codes

Car Parking

Accessible bay

Fig. 6.2 Templeman Library car parking

Fig. 6.3 Platform lift and internal staircases providing access to Level 1

inclusive design. The design solution did not respond to the inclusive design principles that acknowledge difference and diversity by offering flexible features that can be adapted to the users' demands, regardless of their age group, size and abilities. Many interviewees with hidden disabilities avoided using the platform lift, because of the implication that it only catered for people with mobility impairments and wheelchair users. On the other hand, interviewees with mobility impairments were uncomfortable when using the lift as they were exposed to the view of all when using Level 1.

The access audit highlighted the fact that the platform lift did not take account of the needs of all potential users. Its restricted size (980 mm × 1480 mm) could not accommodate more than two users. Moreover, with its push button, it was difficult to operate for people with limited dexterity. Furthermore, users coming from the basement back entrance had to walk a long distance to reach the emergency telephone located in front of the main passenger lift to report when the lift was out of order.

Figure 6.4 illustrates the main barriers to using the platform lift installed in 2010 in the Templeman Library.

On the other hand, interviewees using the back entrance at the basement level were dissatisfied with the quality of the journey they had to undergo to reach Level 1. The back entrance that provided level access was used for delivery and storage purposes in addition to providing access for individuals with disabilities. Many interviewees stated that using the back entrance (Fig. 6.5) resulted in segregation and exclusion, since they had to take different routes from that of their friends in order to access the Library. The poor quality of the journey was made worse by the inconvenient placement of lockers along the back entrance corridor, which created an unfriendly environment. Moreover, bad signage that used inappropriate language, 'disabled lift' rather than 'passenger lift', and poorly placed directional arrows towards the wall instead of the lift created confusion for many users.

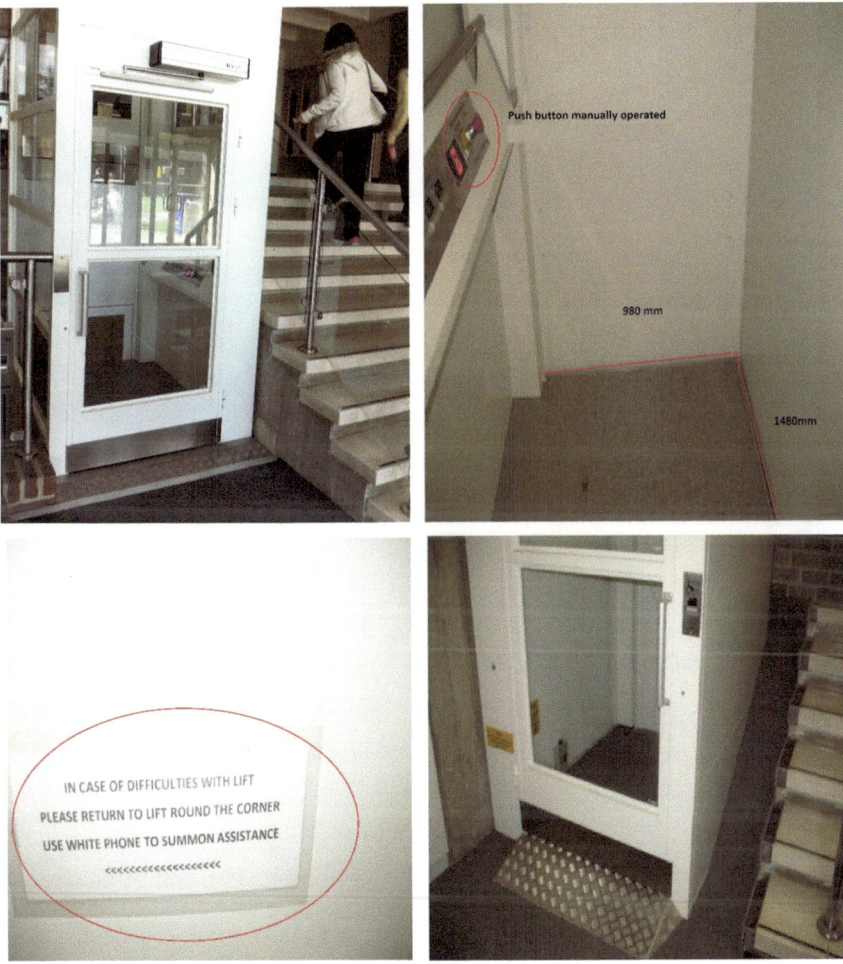

Fig. 6.4 Platform lift and accessibility barriers

Fig. 6.5 Templeman Library back entrance

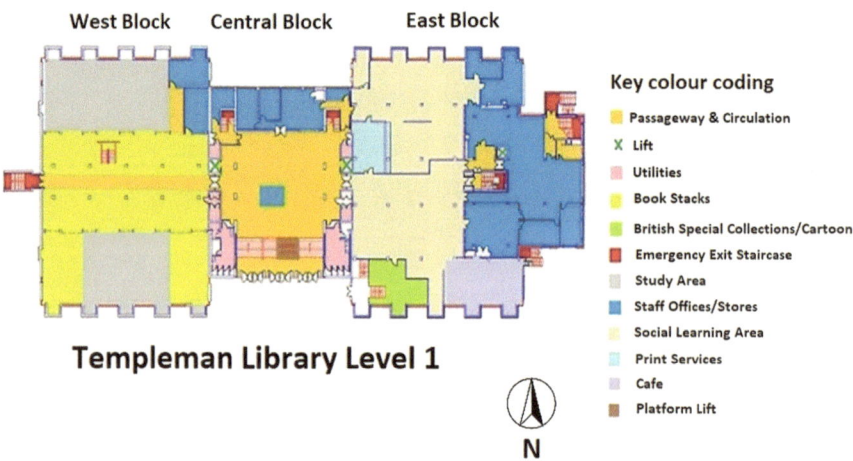

Fig. 6.6 Templeman Library Level 1 plan showing lift locations

In addition to the platform lift, the Templeman Library provides two passenger lifts in its central core and one in its east wing (Fig. 6.6). The two main lifts in the central block and the one in the eastern block at Level 1 resulted in segregation between floors as they allowed for vertical circulation between Levels 2, 3 and 4 but did not provide access to the basement level. The three lifts with their small size (1040 mm × 1400 mm) could only accommodate one wheelchair user. Moreover, the lifts had control buttons that were out of reach for wheelchair users, and people of short stature, whilst the absence of audible announcements and the dim lighting potentially prevented individuals with visual impairments from using the lifts independently.

Another accessibility barrier was that the two main staircases between Levels 1 and 4 were narrow and could accommodate only two users, one ascending and one descending (Fig. 6.7).

A counter provided in Level 1 was an inconvenient distance from the main entrance (around 5 m away) (Fig. 6.8). The counter had a low section for seated users and provided induction loops for people with hearing impairments. However, the turnstiles at the main reception area desk for individuals with disabilities opened towards the user, and this could cause obstruction for individuals with visual impairments or poor balance.

Toilets were provided at the Templeman Library, but they did not cater for all users. Figure 6.9 illustrates the main barriers identified.

The access audit went on to cover the Library management practices and procedures. Real-time observation of students using the Library showed that the main entrance doors were congested, and this could prevent individuals with visual impairments from seeing the power-assisted push pads and the blue visibility enhancements on glazed doors. Moreover, book stacks were obstructed with book trolleys and stools (Fig. 6.10).

Figures 6.11, 6.12, 6.13, 6.14 and 6.15 illustrate the strengths and weaknesses of each floor. Each service is colour coded to highlight the accessibility barriers with the aim of proposing an inclusive design solution that would tackle all these barriers. Figure 6.11 illustrates the basement-level weaknesses, namely narrow and poorly designed passageways, in addition to a central block which serves as a circulation and archive area for special collections, which are not securely protected. On the other hand, the strength of the basement lies in its ability to act as a delivery area close to stores and lifts.

Fig. 6.7 Templeman Library main staircase Level 1

Fig. 6.8 Templeman Library counter Level 1

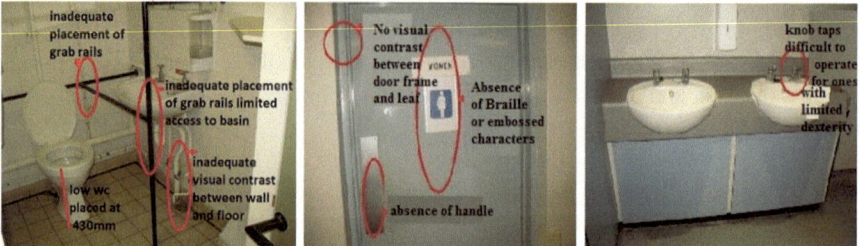

Fig. 6.9 Templeman Library toilet compartments

Figure 6.12 presents the main weaknesses in Level 1, namely that the main entrance does not provide level access and the eastern block does not have containment around its fire escape stairs. On the other hand, Level 1 contains a café, social learning zones and print services in its eastern block close to the main entrance, which could strengthen the area and maximise its usability if it had level access at the main entrance.

Level 2 is considered to be the main group-study space in the building. Whilst the group-study room with its large, bright, high space provides a welcoming and accessible environment for all users, insufficient study desks and narrow spaces between computer desks and carrels in the western block limit the manoeuvring space for wheelchair users. The services provided in Level 2 are not well integrated, with an absence of clear vertical links to library services between central, eastern and northern blocks (Fig. 6.13).

Level 3 includes meeting/training and photographic services. The meeting/training rooms are located in the far corner of the building and are separate from the main services. Moreover, there are narrow spaces between carrels and computer desks, which makes it difficult to manoeuvre wheelchairs. On the other hand, Level 3 with its large, high spaces provides good levels of lighting for all users (Fig. 6.14).

Level 4 contains the special collections, the reading room and the slide collection. The special collections and reading room are located too far from stored material in the basement, whilst the carrels and computer desks again limit manoeuvrability of wheelchairs. The main strength of Level 4 is that it provides a good level of light through its skylights and windows (Fig. 6.15).

Fig. 6.10 Templeman Library management procedures

Fig. 6.11 Templeman Library basement level

6.3.3   Templeman Inclusive Design Proposal

Analysis of the layout and facilities at each level revealed that the central block acts as a link between the western and eastern blocks in addition to providing access via its main and rear entrances to facilitate the use of the utilities, the British Collection and British Cartoon Archive and storage services provided in each of the five levels. The main facilities, such as book stacks, study areas, computer rooms and meeting rooms, are provided in the east and west wings.

It is recognised that the Templeman Library needs to expand to address its shortage of space problem and limited capacity to accommodate facilities for the next

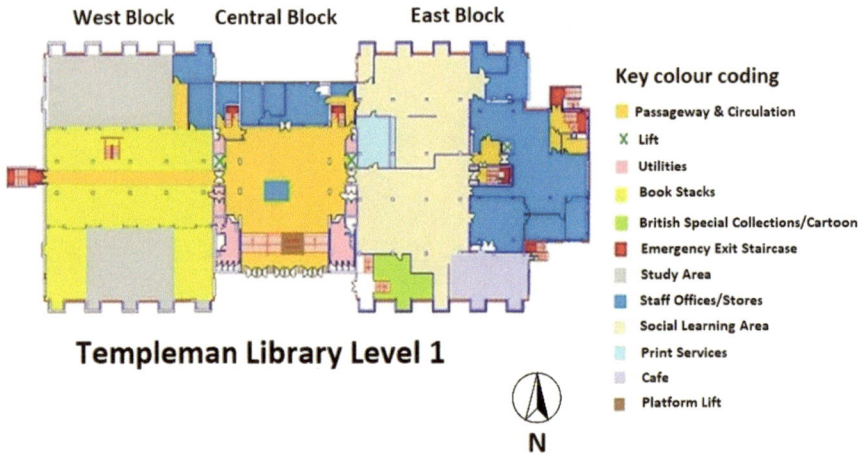

Fig. 6.12 Templeman Library Level 1

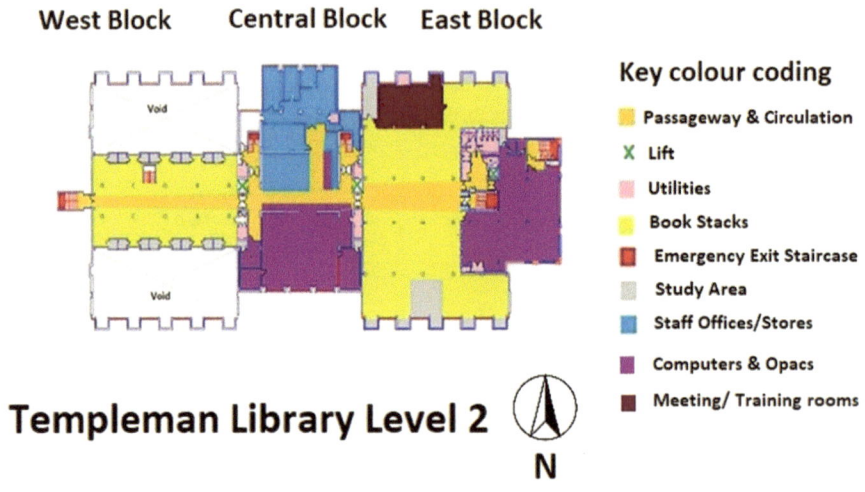

Fig. 6.13 Templeman Library Level 2

15 years, and that it should aim to enhance accessibility and promote inclusion. Hence a proposal, influenced by the architect Shepheard Epstein Hunter's study, to expand the library has suggested remodelling the central block and extending the east and west wings so that the central block becomes the central focal point, provides level access at its main and rear entrances, and bridges the west and east wings. This solution is also evident in the Roundhouse case study presented in the literature review, where a new wing has been created to resolve accessibility barriers.

The Templeman central block at basement level would become the new central glazed atrium, generating a dual entrance by raising the floor by 500 mm and reduc-

West Block   Central Block   East Block

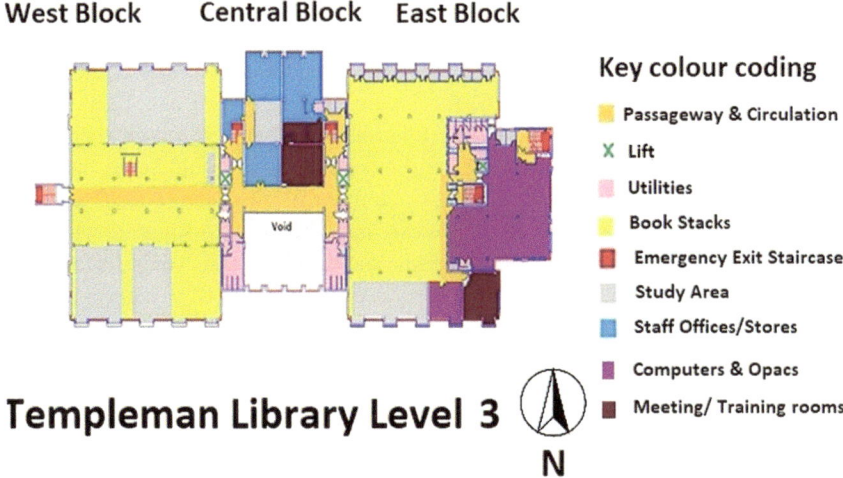

Key colour coding

- Passageway & Circulation
- X Lift
- Utilities
- Book Stacks
- Emergency Exit Staircase
- Study Area
- Staff Offices/Stores
- Computers & Opacs
- Meeting/ Training rooms

Templeman Library Level 3

N

Fig. 6.14 Templeman Library Level 3

West Block   Central Block   East Block

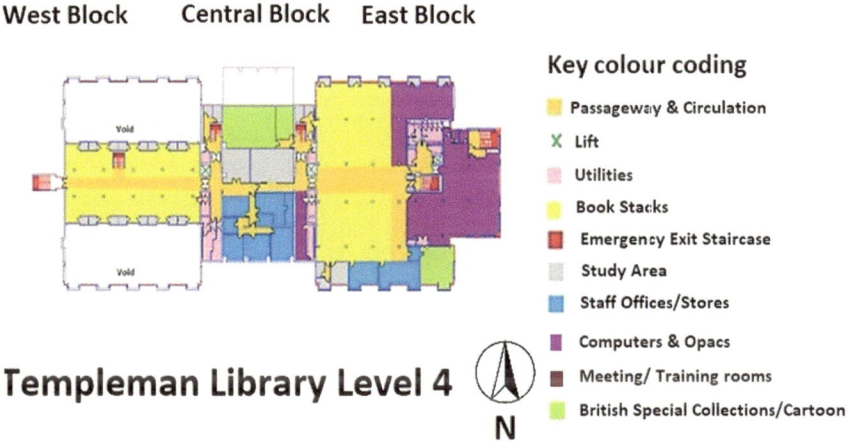

Key colour coding

- Passageway & Circulation
- X Lift
- Utilities
- Book Stacks
- Emergency Exit Staircase
- Study Area
- Staff Offices/Stores
- Computers & Opacs
- Meeting/ Training rooms
- British Special Collections/Cartoon

Templeman Library Level 4

N

Fig. 6.15 Templeman Library Level 4

ing the change in level between the main entrance and the basement, which is currently 1000 mm. The proposal suggests that facilities provided in the central block (British Collection British Cartoon Archive, staff and storage services) be relocated to the extended part of the east and west wings whilst keeping the sanitary facilities, lifts and emergency exit staircase in the same block. The proposal also suggests expanding the two lift shafts so each passenger lift can accommodate two wheelchair users in addition to a few standing individuals. The proposed lifts, 2000 mm width × 1400 mm length, would comply with the Part M Building Regulation. Call and control buttons would be within the reach of wheelchair users; they would have Braille characters, audible announcements and handrails along three sides. In addition, they would incorporate a mirror, walls with adequate visual contrast that would not cause reflection or glare, and good lighting (Fig. 6.17). Moreover, the proposal suggests relocating the passenger lift to the eastern wing as well as providing a new passenger lift in that wing, when the eastern and western blocks have been extended.

The central atrium would provide a bridge on Levels 1, 2, 3 and 4 linking the eastern and western blocks, whilst one internal staircase is proposed to provide vertical circulation between the basement and Level 1. It would serve the main entrance and provide views overlooking Canterbury. Whilst the proposal suggests maintaining the existing locations of fire exit staircases, it also suggests widening the staircase (1600 mm width, 150 mm riser) and upgrading the handrails and nosings on steps so they are clearly visible to all users. The proposal to incorporate the existing staircases, along with the new proposed staircase, bridges and two passenger lifts, would provide alternative vertical and horizontal circulation options for users. The two glazed main entrances could be easily accessed by providing automatic doors, which would be clearly marked at two heights and clearly visible against their backgrounds (Fig. 6.18). Moreover, it is proposed to relocate the reception desk so it would face the visitors coming from the two main entrances. It is suggested that the desk should serve all users by providing high and low counters as well as a hearing enhancement system.

To identify the services and facilities provided at each level, a colour coding system is recommended so that a subtle pink would indicate the basement level, yellow Level 1, blue Level 2, brown Level 3 and orange Level 4. Whilst these colours would be used to highlight the facilities provided on each floor and each block, adequate signage with appropriate visual contrast would be employed to indicate each level and the facilities provided on each floor. In addition, colour-coded floor plans would be placed in the central atrium at each level. It is recommended that the colour-coded signage with information about the services provided at each level should be incorporated in smartphone applications that users could download and gain access to. The proposal also suggests providing bookshelves for periodicals and frequently used reference materials at a range of heights (400–1300 mm) within the reach of wheelchair users, and short people. Book stacks are usually designed to store compact library materials and thus most are not within the reach of wheelchair users so the proposal recommends that staff members should be available to help. Moreover, all bookshelves and book stack aisles should be 1100 mm wide, thus allowing adequate manoeuvring space for wheelchairs. It is important to ensure that adequate lighting is used in the vicinity of the bookshelves to maintain the levels of illumination for individuals with visual impairments. In addition, the proposal suggests using a colour coding system when labelling books according to subject. The proposal recommends providing computer catalogue systems placed at 750–850 mm from floor level, and at least one computer terminal with software which would enlarge text for people with visual impairments. It is recommended that once this facility is in place, it should be clearly signed. Also, the proposal suggests providing accessible reading desks which would have a height of 750–850 mm from floor level, and the distance between one row and another should be 1400 mm, which would allow space for wheelchair users to manoeuvre. Task lighting should be installed over each reading desk.

The design proposal in Fig. 6.20 replaces an improvised space, which has resulted in exclusion and segregation of users, with open, neat and integrated spaces, which would offer dignity, free choice and equal opportunity for all users. The dual, easy-gradient entrances would give users the choice to access the library from either the main or the back entrances. Both entrances would have automatic doors that would slide open from either direction, thus benefiting wheelchair users, people using canes or crutches, people with visual impairments, students carrying books or people with any other limitations. A reception desk with a high and a low counter and embedded induction loops would benefit seated users or users short in stature, and would allow people with hearing impairments to communicate with reception-ists, whilst high desks would benefit standing users. A colour coding system on each floor providing information about each floor surface, with room numbers written in Braille and using embossed characters and pictograms, would help all users by clearly showing the facilities available on each floor. Moreover, wide passenger lifts, internal staircases and bridges would offer the user a choice of different vertical and horizontal means of circulation, allowing them to move easily from one level or facility to another. Adjustable desks, low shelves and furniture with different seating types would allow users to reach books, adjust the desks and choose seating to accommodate their specific needs. Moreover, sanitary facilities with accessible toi-lets, standard toilets and baby changing facilities would offer users the chance to use these facilities without difficulty.

6.4 Eliot College

6.4.1 Overview

The College is named after T. S Eliot, the poet, who died on January 4, 1965, the same day the University of Kent was formally established. Eliot College was the first to be constructed at the university. The basic design was inspired by Louis Kahn's design for a residential block at Bryn Mawr College in Pennsylvania (Martin 1990). The College was designed as several square blocks, accommodating 300 study bedrooms and other services and facilities. Each square unit or block contains distinctive interior spaces with study bedrooms along all four walls. The services and facilities include refreshment areas, such as Mungo's and Eliot Hall, music practice rooms, and accommodation and offices that are located at Level 2. Level 3 contains the main reception area, drama rehearsal rooms, lecture theatres, a snooker room, the Eliot shop and storage facilities. Level 4 includes staff common rooms, music master room, and the Music Society office and seminar rooms. Eliot College is open 24 h per day, but its reception area, located at Level 3, opens from 7:30 am till 8:00 pm, Monday to Friday, and 8:00 am till 4:00 pm at weekends.

Students and visitors can reach Eliot College either by using the bus drop-off services provided at Keynes and Darwin Colleges or by car parking space being provided near the Mandela Building. There are two designated accessible parking bays in the car park (Fig. 6.16).

Eliot College Car Parking

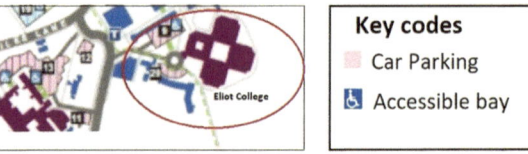

Fig. 6.16   Eliot College car parking

6.4.2   Eliot College Accessibility Barriers

Built in the early 1960s, Eliot College represents one of the university buildings that were not well designed to accommodate users with disabilities. Whilst the University of Kent at that time offered services to individuals with visual impairments by providing tape-recorded books, it did not focus on eliminating physical barriers that restricted many individuals with different types of disability from accessing the built environment independently. In 1965 Ann Smith became the first wheelchair user to be enrolled at the University of Kent after being accepted without the normal interview (Martin 1990). To accommodate her specific needs, grab rails were provided in her study bedroom at Eliot College; however, the university could not remove all the architectural barriers, such as steps and different levels in Eliot College and Rutherford College and the high shelves at the Templeman Library. To overcome such barriers, Ann used to seek assistance from her colleagues, who had to carry her and her wheelchair to negotiate a change in level. Ann's accessibility experience influenced the university's decision-makers when it came to enhancing accessibility for people with mobility impairments at new and existing buildings rather than catering for a range of different disabilities. The notion that disability is associated with mobility impairments, particularly the needs of wheelchair users, is reflected in the current specific provisions (ramp and lift) provided at Eliot College, and in the failure to resolve the orientation difficulties and barriers that affect most people inside the college, including people with sensory and cognitive impairments.

The Eliot College main entrance results in segregation of its users. It is located at Level 3, and can be accessed by using either the steps or a ramp which runs alongside. It can also be accessed from the car park facing Level 2 by using two external staircases (Fig. 6.17).

Although the main entrance has a ramp, it is difficult to manage due to its narrow width, long and steep gradient and high handrails. In addition, a mesh-finish balustrade along the handrail creates confusing shadows for individuals with visual impairments. On the other hand, the external steps at Levels 2 and 3 are problematic for many users, especially those with visual impairments, who struggle to manage the inconsistencies of step risers as well as the absence of nosings and lack of tactile warnings on the top and bottom of steps.

Orientation, mapping and route-finding are the common barriers that most interviewees highlighted when navigating inside Eliot College. The absence of

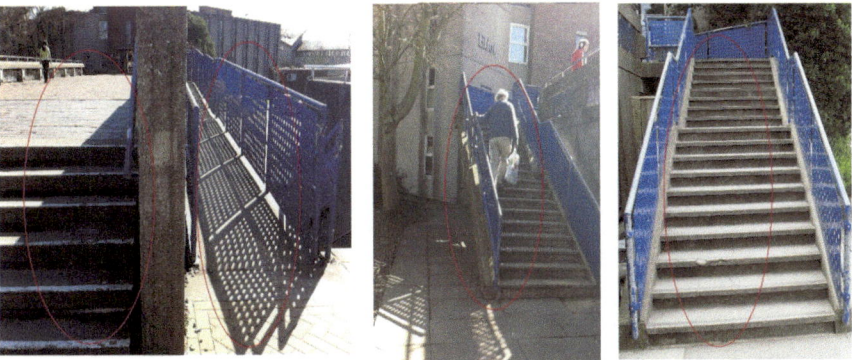

Fig. 6.17 Eliot College main entrance pedestrian routes

Fig. 6.18 Floor finish in east and north wings

directional and informational signage along corridors and the spatial configuration of the building cause disorientation and discomfort for many users. Although efforts have been made to improve orientation by differentiating the floor finishes in each wing, there is no information about the colour notations used in each floor or wing. Figure 6.18 provides an example of the floor colours used in the north wing, which has a red vinyl floor, and the east wing, which has a green finish. Moreover, the reflective finishes create glare for people with visual impairments.

Although a passenger lift close to the reception area in the north wing was installed to provide vertical circulation between all floors, many interviewed individuals with disabilities coming from other wings had to walk a relatively long distance to reach it and found this tiring. Moreover, many found difficulty in identifying the lift location in the absence of signage along the corridors and passageways (Fig. 6.19).

Figures 6.20, 6.21 and 6.22 illustrate the strengths and weaknesses of each floor. Each service is colour coded to highlight the accessibility barriers, which require an

Fig. 6.19 Eliot College passenger lift

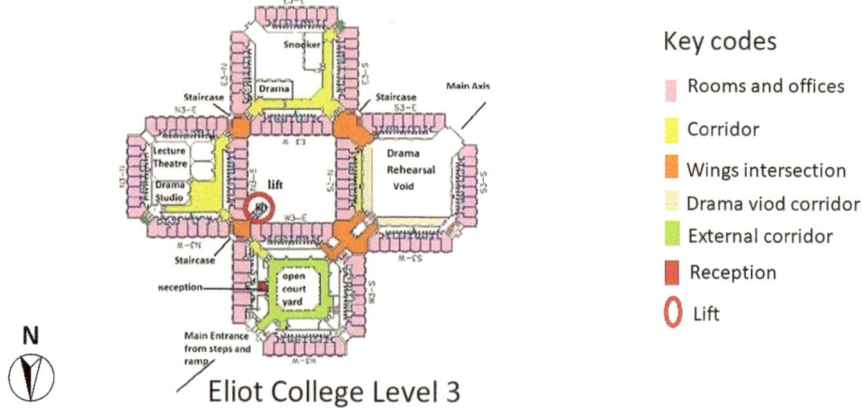

Fig. 6.20 Eliot College Level 3

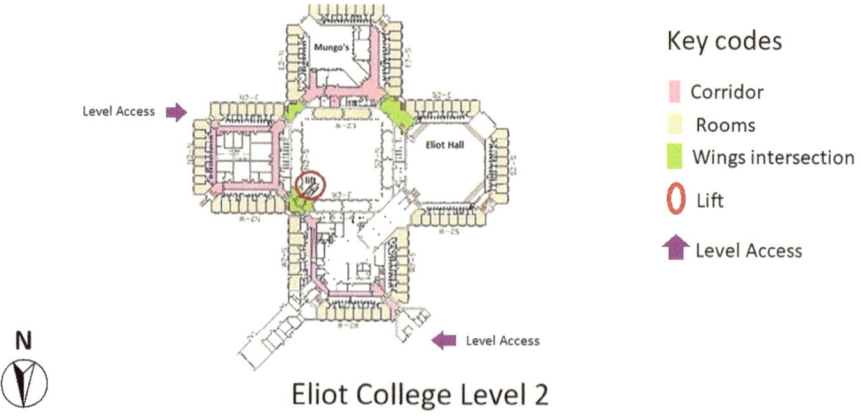

Fig. 6.21 Eliot College Level 2

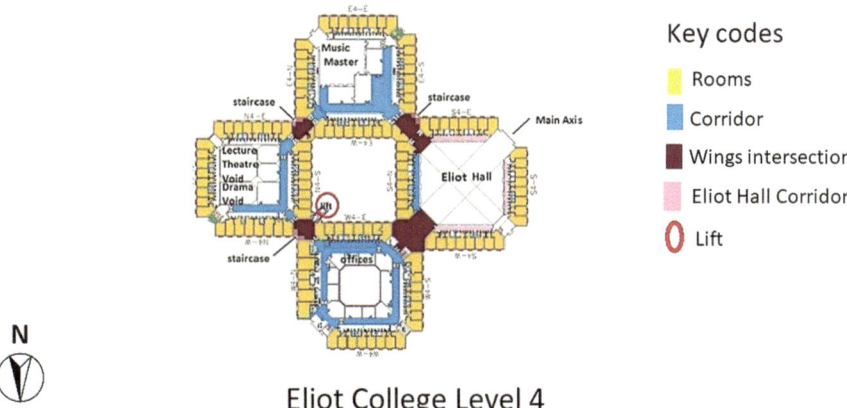

Eliot College Level 4

Fig. 6.22 Eliot College Level 4

inclusive design solution to tackle them. Figure 6.20 illustrates the Level 3 weaknesses arising from the location of the reception area, which is isolated from the main services. Moreover, providing one passenger lift in the north wing some distance away from other wing entrances extends the journey for users coming from the west and south wings, and causes disorientation for many users, especially individuals with visual and cognitive impairments. An open courtyard, close to the main entrance, could act as a main axis route or focal point that would remove the orientation barrier.

Figure 6.20 illustrates the facilities provided at Level 2, which include offices and accommodation facilities in the north wing, a refreshment area (Mungo's) in the east wing and a multifunctional facility (Eliot Hall) in the south wing. The main weakness in Level 2 is the absence of level access to Eliot Hall, which is resolved by providing a platform lift that does not comply with Part M. On the other hand, Level 2 provides two level accesses to its north-east and south-west sides, which, if clearly signposted, could act as main entrances. Level 4 (Fig. 6.21) includes staff common rooms, seminar rooms and office and music facilities such as music master room and Music Society office, in addition to accommodation. The main weakness in all the three levels is the lift location that is far away from many of the facilities. On the other hand, Level 4, with its large, high spaces, provides good levels of lighting for all users.

6.4.3 Eliot College Inclusive Design Proposal

The access audit, feedback from users and detailed study of the floor plans of Eliot College revealed that although some provisions and features do not cater for all users, the building has the potential to create an inclusive and friendly environment.

This could be achieved by firstly enhancing the external routes that lead to the college so that they can be used by a wide range of people. Thus the proposal replaces the main stepped entrance and ramp at Level 3 by a gently sloping passageway or bridge (1/30 gradient) that would link the college to the main campus and other surrounding buildings. It is recommended that the passageway should have two handrails that would contrast visually with the surrounding surfaces. The proposal suggests that the passageway should be clearly signposted once it has become a main entrance to the building. Figure 6.23 shows the design proposal. This suggests that all external staircases should be upgraded to non-slip step surfaces with even risers that have well-contrasting nosings and corduroy tactile warnings at the top and bottom of steps. It is recommended that the staircases should have handrails running along the top and bottom of steps and landings, and that they should contrast with the background. The flexibility and free choice for potential users could be exemplified by making the two level entrances at Level 2 extra main entrances in addition to the one at Level 3. All three main entrances should be clearly signposted.

To tackle the disorientation problem at Eliot College, it is recommended that all floors adopt a colour coding system, such as the one used in the Carrington Building at the University of Reading, when redecorating the building. Once the colour coding system has been adopted, it is recommended that a tactile map of each floor should be clearly signposted. The maps should include information about the colour coding system and facilities provided on each floor, and should be placed at two heights suitable for standing and seated users. Such maps, together with a virtual map application that could be downloaded on users' smartphones, would give the users choice when it came to accessing such valuable information and would help them plan their journey or route.

Fig. 6.23 Entrance proposal—not drawn to scale

The proposal also suggests that a new reception area should be provided at Level 3 in the open courtyard area so that it would become the focal point of the building. Toilets with one accessible WC and a family compartment with changing units for mothers could be provided in the relocated reception area. It is also recommended that at least one passenger lift should be installed in the south wing to accommodate the number of students and staff members using Eliot College and which could be used as an emergency evacuation lift.

Inclusiveness should be reflected not only inside the above-mentioned buildings, but also in the ways users interact with their surroundings. This should include the means of transportation used; for example, university buses should have level access, and clearly contrasting material finishes to aid individuals with visual impairments. All university buses serving the university should have audiovisual announcements to alert users with visual and hearing impairments as they approach the desired drop-off points. Providing an inclusive means of transportation at the university is vital to promoting integration. This has been noted in the case study of the Eden Project gardens, where accessible and family parking bays near the main entrances are provided, as well as accessible shuttle buses that transfer users from one site to another. Thus accessible university buses with similarly accessible bus stops that are equipped with a canopy and a seating area, which can be used by a broad spectrum of users, are vital. The bus stop area should also display either an A3 paper timetable, with text that is in 14 pt Arial Bold type (RNIB 2014), or an interactive screen that has both visual and audible announcements to suit different users.

Directional and information signage is an important design feature that both universities should provide along external routes to indicate building directions. The information should be provided in text using a large font, embossed characters and Braille. Providing signage with pictograms, such as the ones proposed at the Willows School and the Eden Project gardens, illustrated in Chap. 3, would enable first-time visitors and potential users to find their way around buildings.

It is recommended that a tactile site plan of the university with embossed text and Braille to indicate the building types should be provided and also a colour-coded system of the services within the buildings.

External routes leading to the above-mentioned four case studies and other university buildings should have tactile blister paving on dropped kerbs to assist people with visual impairments to recognise the position of such kerbs. Moreover, external routes should be regularly maintained so they are level and do not present trip hazards for people with mobility and visual impairments. Carefully selected external furniture, such as that provided at the Eden Project gardens, that anticipates the needs of a wide range of users should be installed in the outdoor built environment in the two campuses. By anticipating users' activities and needs in both the external and internal environments in university settings, both universities could achieve inclusiveness in their built environments.

6.5 Concluding Summary

The form of the built environment represented in the above-mentioned institutional buildings of the four universities reflects the professional values and practices of the twentieth century. These were centred on the investment available and suppliers' decisions rather than users' requirements. Such an approach produced exclusive environments that were insensitive to potential users' needs and prevented individuals with disabilities from having equal opportunities to interact with their societies. Buildings that have stepped entrances and those that provide a long and steep ramp alongside a stepped entrance, examples of which are both to be found at Eliot College, and those with platform lift that can only be used by wheelchair users, as in the Templeman Library, are all instances of design features that are not appropriate for many users, including people with visual, hearing and physical impairments.

The creation of inclusive environments is becoming a priority in the twenty-first century with the growing awareness of disability rights and anti-discrimination legislation, which aims to integrate individuals with disabilities into the mainstream society. The University of Kent nowadays has a legal duty to develop its built environment to accommodate the needs of individuals with disabilities.

Findings from the study revealed that although building regulations have influenced the accessibility level of university buildings, they have failed to create spaces accessible to a broad spectrum of users. Designing beyond Building Regulation compliance can create spaces that are visitable, comfortable and enjoyable for all potential users. To achieve inclusive spaces that accommodate a broader range of human capabilities in order to promote social inclusion and cultural respect, a great deal of thought about potential users' needs and how to involve them in the design process is vital in the design and implementation phases. Such inclusion is reflected in the two design proposals. Each case study, by responding to different accessibility barriers (level access, orientation, signage and visual contrast), reflects the diversity of people's needs by offering freedom of choice, flexibility and comfort. The advantage of inclusive design reflected in these two case studies is that it responds to the needs of the people interviewed during the data collection. Acknowledging that the main and back entrances at the Templeman Library were problematic for most interviewed users, the inclusive design proposal tackles that issue by offering choices, namely using either the main front level-access entrance or the main back level-access one, in addition to having the freedom to choose to use the internal staircases or the passenger lifts provided at the main entrance. Another example of the importance of responding to the users who were interviewed and surveyed in order to create an inclusive environment is reflected in the case of Eliot College where many respondents found it difficult to find their way around. Adopting a colour coding system at each level and providing maps in different formats, that is, printed maps, tactile maps, directional and information signage with information about the colour coding system and facilities provided on each floor, placed at two heights suitable for standing and seated users, can benefit all people and provide options for most of them when navigating their way around the building. By using knowledge acquired from interviewing the users of buildings such as those which were the subjects of the two case studies, design can better respond to people's needs and wants, and that is one of the main advantages of inclusive design.

Another important advantage revealed by these case studies is that it can provide features that are easy to use and understand. Automatic entrance doors, such as those proposed at the Templeman Library, are a feature that is easy to use by everyone, whether they are short or tall, have limited dexterity, are students holding books, etc. Moreover, inclusive design provides environments that are adjustable and can be used by people of different ages, sizes and abilities. The availability of adjustable workstations and desks at the Templeman Library, for example, can satisfy a variety of needs.

An important advantage of inclusive design is that it goes beyond the minimum standards of Building Regulations to provide access for a broad range of people. An example of that is the provision of gradual inclines in the approach to Eliot College rather than the stepped entrance to replace the ramp that was provided earlier. Such gradual inclines as those proposed at Eliot College can benefit all people, including wheelchair users, those carrying luggage or staff members using trolleys. Moreover, inclusive design addresses aspects that promote health and safety for users. The use of non-slip and non-reflective finishes within the passageways in Eliot College, for example, and toilet floor finishes are designed to ensure security for users of such facilities. Furthermore, providing audiovisual emergency call systems in passenger lifts and inside accessible toilets, together with emergency exit routes and refuge areas, will ensure that people can be rescued in a timely manner in case of emergencies. For all of these reasons, inclusive design goes beyond the minimum requirement for accessibility as it aims to create products and environments that are adaptable, flexible and secure for the people using them.

References

Building Control Guidance. (2016). *Building control guidance note*. Retrieved July 20, 2018, from https://www.charnwood.gov.uk/pages/building_control_guidance_sheets

Burton, E., & Mitchell, L. (2007). *Inclusive urban design: Streets for life*. Oxford: Architectural Press.

Equality Act. (2010). London: The Stationery Office.

Martin, G. (1990). *From vision to reality: The making of the University of Kent at Canterbury*. Canterbury: University of Kent at Canterbury.

RNIB. (2014). *How do I create accessible forms?* Retrieved July 30, 2018, from https://www.rnib.org.uk/search/site/Papers%20in%20accessible%20formats

Sawyer, A., & Bright, K. (2007). *The access manual: Auditing and managing inclusive built environments*. Oxford: Blackwell Publishing Inc..

Chapter 7
Barriers to Inclusive Design at University Built Environment

7.1 Introduction

This chapter describes and evaluates the findings of the study and its implications for the creation of built environments that people can enjoy and use independently and safely at a university built environment.

More specifically, it answers the general and specific research questions presented in Chap. 4 and re-presents in this section as follows:

1. What are the physical and management barriers which hinder individuals with disabilities from accessing the university?
2. To what extent does the University of Kent, Canterbury, comply with the Building Regulations to influence access and eliminate architectural barriers?
3. To what extent are students, with or without disabilities, and staff members satisfied with the services provided?
4. Which inclusive/accessible approach would users most like to see adopted at the University of Kent, Canterbury?
5. To what extent have education providers managed to make reasonable adjustments or eliminate architectural barriers?
6. To what extent are architects aware of the needs of potential users and people with disabilities when designing university buildings?
7. To what extent are architects aware of the inclusive design approach?
8. To what extent do architects abide by the design rules and regulations in their designs to cater for all users, including people with disabilities?
9. Has the guidance influenced architects' designs of new buildings or limited their creativity?

Five main barriers to inclusive design emerged from the analysis of the two research phases. These are (1) sociocultural differences and inclusive design, (2) failure to define inclusive design and disability, (3) accessible design and regulation barriers, (4) procedural barriers and (5) organisational barriers.

© Springer Nature Switzerland AG 2020
I. Shuayb, *Inclusive University Built Environments*,
https://doi.org/10.1007/978-3-030-35861-7_7

7.1.1 Sociocultural Differences and Inclusive Design

The results of this study explain the different roles of cultural and economic values, as well as the attitudes towards users, particularly individuals with disabilities, and how they influenced the practices of architects and designers responsible for shaping the university's built environment. The UK is an example of a developed country and cultural attitudes towards individuals with disabilities were reflected in the views of architects who were interviewed at the University of Kent. They acknowledged their legal and moral obligations to eliminate architectural barriers for individuals with disabilities by complying with Building Regulations, such as Approved Document M and British Standards. Although these regulations enhance accessibility for individuals with disabilities, they have resulted in marginalised and segregated university facilities and services that are biased. This finding is also noted in the University of Arizona case study presented in Chap. 3, which revealed that both existing and new buildings failed to anticipate the needs of potential users.

Negative cultural attitudes towards individuals with disabilities and lack of knowledge about the needs of different types of disability must be addressed and changed. A different approach to designing university built environments should be acknowledged. Architects shaping university built environments should be required to adopt the inclusive approach that recognises the diversity of their users' lifestyles and uses their individual experiences to identify and remove the physical barriers created and designed for able-bodied lifestyles, attitudes and culture. Changing the attitudes, values and practices of architects in response to the needs of all users of their buildings (Steinfeld and Maisel 2012; Nussbaumer 2012; Burton and Mitchell 2007; Imrie and Hall 2001) is vital when it comes to achieving inclusive university environments.

This new approach that cares about human social lifestyles and healthcare issues is supported by Finkelstein (2002) and Swain and French (2000) who propose the restructuring of society to make it more diverse. This approach rejects the cultural attitudes and assumptions that individuals with disabilities have suffered personal tragedies, and casts aside the social model that sees the problem as one that exists within society. The inclusive approach that this book has adopted aims to project architecture as a tool to be used to achieve social justice. It aims to integrate all services to cover the needs and desires of the whole population and provide spaces and provisions that can improve the quality of life by promoting a healthy environment and better physical environments and workplaces. The design strategies illustrated in Chap. 6 are all good examples of those that can achieve inclusiveness for a wide spectrum of users.

If they are to acknowledge users' lifestyles, their stage of life and the changes that take place as people grow older, architects in charge of shaping university built environments should place emphasis on functionality, user-friendliness and simplicity without making compromises with regard to aesthetics and functionality. Through awareness campaigns individuals with disabilities could present their own personal experiences and, in this way, instruct and educate architects and designers so that they might develop a better understanding of users' needs.

As well as requiring university architectural programmes to incorporate inclusive design, education could also help to eliminate negative cultural attitudes. The University of Kent, as well as other universities in the UK, should involve the students and academic staff in the design and the decision-making processes. These individuals should even be consulted about existing buildings. They could play an essential part in shaping the built environment of their university because they would be able to address their own concerns, which are frequently overlooked by architects.

7.1.2 Misinterpretation of Inclusive Design and Disability

To encourage the implementation of inclusive design in university built environments, this book aims to demonstrate what has obstructed the adoption of the inclusive design approach at the University of Kent.

One of the barriers to achieving inclusive design was the misinterpretation of the concept. The research findings reveal that the architects who were interviewed associated inclusive design with design that anticipates the needs of individuals with physical disabilities, thus enabling them to access university built environments. As Steinfeld (1994) argues, enhancing accessibility for individuals with disabilities aims to allow them to use accessible environments; however, this does not respond to the diverse needs and social inclusion of all users, which inclusive design does acknowledge.

An interesting finding from the population distribution at the University of Kent reveals that the demographic trend at the university is an important factor when it comes to understanding the scope of inclusive design. Architects and designers can take advantage of a business opportunity to apply their innovative solutions to tackling these diversities in university buildings if they anticipate these demographic trends. As Steinfeld and Maisel (2012) state, a better understanding of demographic trends and changes can enhance architects' understanding of the purpose of inclusive design and its impact on predicting future market demands.

As inclusive design is centred on accomplishing social justice and developing individual capabilities, universities aiming to achieve inclusiveness should draw attention to the different and diverse needs of potential users. By acknowledging their diverse cultural lifestyles, religion and experiences, university built environments can respond to these needs.

Provision of more sanitary units for females, baby changing areas, family toilets and day care for children, as well as accommodation facilities for couples, is recommended at universities in order to take account of the needs of females as findings have revealed that the number of women and girls is around double the number of males at the University of Kent. Moreover, whilst the major proportion of the population at the University of Kent is young (between 18 and 25 years old), there are around 6% (between 50 and 65 years old) who are likely to be more impacted by physical barriers as they get older.

Around 23% of the University of Kent population have disabilities, with approximately 18% having temporary disabilities, whilst only 5% have permanent disabilities. Moreover findings concerning the statistical distribution of the population at this university reveal that around 45% of the population have cognitive impairments and learning difficulties/dyslexia. By acknowledging factors associated with the health and wellness factors that can limit users' ability to manage the built environment and the increase in the age of retirement, universities aiming to practise inclusive design should be required to respond to the future demands of such users by providing inclusive built environments from which people from different age groups and with different abilities can benefit. Studying the population distribution at universities and its variables within age groups, nationalities, sexes, religious practices and disabilities would highlight the prominence that architects and designers should accord to these variables in order to identify the needs of the different groups and their social diversities, which would lead to a better understanding of the purpose of inclusive design and would generate insight into how to produce an inclusive university environment.

Moreover, findings show that there is no consensus about the definition of disability among the stakeholders interviewed. Whilst most stakeholders at the University of Kent recognised disability as an impairment that restricts an individual's ability to interact with the built environment, they only acknowledged three types of impairment: mobility, visual and hearing impairments. Stakeholders categorised individuals with disabilities as people with impaired mobility who would be impacted by physical barriers to the greatest extent, and ignored other groups who might be affected by the same barriers. The other groups include people with mental and cognitive impairments, heavily pregnant women, obese people, those of short and tall stature, mothers/carers, children and elderly people.

Interviews with architects suggested that many of them adhered to the medical model of disability in responding to the design needs of individuals with disabilities, relating impairments to medical conditions that are mainly associated with mobility deficiency. This illustrates the narrow understanding of disability that responds to wheelchair users' needs rather than acknowledging a wide spectrum of disabilities and other groups of users who might be impacted by the same barriers. The access audits at the six buildings at the University of Kent confirmed this finding, which is manifested by the provision of ramps set off to the sides of main- or back-entrance stairs or platform lifts alongside internal stairs, but the failure to tackle other users' needs, such as those of older people, and women, men, young children and people with other disabilities. Inclusive design provides a design solution that tackles all of these diversities by, for example, providing level-access entrances to which all can have equal access, or a passenger lift that is clearly signposted with tactile signage, has visual contrast between its material finishes, has Braille or embossed call buttons placed at a reachable height for both seated and standing users, and provides audible announcements, adequate lighting and an emergency call button with easy-to-read information, all of which could accommodate the diverse needs of potential users. This common finding, noted in other research (Burton and Mitchell 2007; Imrie and Hall 2001), highlights the fact that the built environment continues to meet only the needs of people with physical impairments, particularly wheelchair users, focusing on the disability rather than on the diverse lifestyles and experiences of potential users.

Whilst the misinterpretation of inclusive design and the medical definition of disability were shown clearly in the current research by the association of disability with mobility impairments and in particular the needs of wheelchair users, the students' disability classifications at the University of Kent for example show that learning difficulties (dyslexia), hidden disabilities and mental health difficulties form the greater proportion of disability types (Table 7.1).

A key finding was that professionals involved in shaping the built environment tended to categorise individuals with disabilities as people with impaired mobility, without acknowledging other disabilities, namely learning difficulties, mental health difficulties and hidden disabilities, a blinkered approach which has led to exclusion. This exclusion was reflected in one architect's response to defining disability.

> An individual with disability is a person who has mobility impairment or a person who uses a wheelchair that restricts him/her from accessing the built environment. In our profession we only cater for people with physical impairments and the elderly and do not consider the needs of people with visual or hearing impairments because we believe that they can walk and access the built environment.

A key finding from interviews with architects was that they and the designers in charge of creating built environments at the University of Kent lacked knowledge about users' and occupants' needs, which had resulted in the creation of accessible environments that do not tackle all users' needs.

To create an inclusively designed environment in a university setting, architects and designers have to understand the relationship between architecture and inclusive design. It is necessary for them to study the demographic trends and their interaction with social and cultural backgrounds in order to accommodate people's needs according to their various lifestyles.

The concept of inclusive design and its relationship with disability should be well defined and understood not in terms of impairments, but with respect to the social, cultural and physical issues impacting, and interfacing with, the world in which all potential users live. This redefinition is supported by findings from the

Table 7.1 University of Kent—four most common disability types based on the DDSS statistics

	2007–2008		2008–2009		2009–2010		2010–2011		2011–2012	
Disability	No.	%	No.	%	No.	%	No.	%	No.	%
Learning difficulties	566	46.1	686	46.4	686	46.4	736	45	785	44.2
Unseen disabilities	233	19	230	15.6	230	15.6	239	14.6	230	15.6
Mental health	61	5	126	8.5	126	8.5	189	11.7	270	15.4
Mobility difficulties	61	5.0	48	3.2	48	3.2	60	3.7	74	4.3

literature (Woodhams and Corby 2003; Gooding 1996; Swain and French 2000) and interviews with individuals with disabilities who call for their inclusion in the mainstream of society in order to have access to better environments and workplaces that promote equality and diversity, and to have their differences accepted. Clearly, a better understanding of the definition of inclusive design would help in clarifying its practical purpose, which is centred on achieving social inclusion, equality and independence for a broader population than just those individuals with specific types of disability.

To achieve an inclusive university environment, architects and designers should gather information about their users' lifestyles, experiences and diversities in relation to using the built environment. Hence, there should be collaboration between architects, designers and users, with designers acknowledging the users' capabilities to use the environments and services, and identifying the physical, social, cultural, emotional, sensory and cognitive factors that affect the interaction of users with the environment or service.

Feedback from users in the study highlighted the essential requirement to involve them during the design phase. Most users blamed the existence of accessibility barriers at the University of Kent on their exclusion from the design and implementation phases. One participant with mobility impairment noted that the University of Kent never consulted users when refurbishing a building, and although he had volunteered in writing to propose suggestions for the enhancement of accessibility, these suggestions had not been seriously taken into consideration by the Estates Department at the University. Although most education providers recognised the importance of users' participation in eliminating barriers and creating an inclusive environment, they admitted that such involvement did not take place. One education provider noted that accessibility issues had arisen as soon as a new building opened due to lack of consultation with users.

The collaboration between designers and end users should occur at all stages of the design process from the pre-planning stage and throughout the construction phase to attain an environment or facility that addresses the needs and aspirations of a wide range of users, including marginalised groups and individuals with disabilities. Projects based on the participatory approach reviewed in Chap. 3, that is, the Carrington Building in the University of Reading, the Willows School in Wolverhampton, the Eden Project gardens and the Roundhouse, exemplify the architects' and designers' understanding of the users' lifestyles and cultures, which enabled them to create inclusive environments from which all people regardless of their age, culture or ability could benefit.

7.1.3 Accessible Design and Regulation Barriers

Difficulties in understanding the difference between accessible design and inclusive design are reflected in the built environment at the University of Kent. Findings highlight the fact that architects and designers saw inclusive design as that which complied with accessible regulation standards to provide accessible designs and special features, which tended to promote exclusion and social isolation.

Accessible design offers separate facilities for people with disabilities, which lead to exclusion and seclusion (Goldsmith 2001; Steinfeld 1994; Steinfeld and Maisel 2012; Nussbaumer 2012). The ramp provided in Eliot College set off to the side of the staircase, the passenger and platform lifts provided in the Templeman Library and Eliot College, and the Templeman Library back entrance are all examples of accessible design. These examples of accessibility are underpinned by the medical model of disability that promotes a negative and stigmatising view of individuals with disabilities. A significant limitation of accessible design was highlighted in the views of individuals with disabilities who were interviewed at the University of Kent, describing these specific features as unreliable, expensive and difficult to repair. Not only did these specific features hinder people with disabilities from accessing the built environment, but they also presented a negative image of such individuals with impairments and disabilities. The dominance of the medical model of disability reflected in such specific provisions and the association of accessibility with impairments without acknowledging the needs of other potential users, such as women, men, young/old people, those of short/tall stature, and right- and left-handed individuals, promoted an exclusive university environment that did not envisage social justice, equity or diversity of its users' needs.

Feedback from architects revealed that most of them relied in the first place on Building Regulation standards to enhance the level of accessibility for individuals with disabilities at universities. Architects at the University of Kent used legislative building standards, such as Approved Document M and British Standards, when extending an existing building or designing a new one and catered only for physical disabilities. Although these regulations assist architects in enhancing the level of accessibility at some university buildings, findings from interviews with architects and access audit assessments revealed that these regulations alone are an insufficient tool when it comes to achieving inclusive built environments as they set minimum standards based on accessible design. Such standards concentrate on covering the access needs of individuals with mobility, visual and hearing impairments and do not provide access and egress provision for other users, such as people with cognitive impairments, temporary disabilities, dyslexia and asthma, who may experience difficulties in accessing or evacuating university buildings.

> The type of disability should be better described in the Building Regulations and, being an architect, I think that the standards should try and find a kind of intermediate, or a point where one tries to accommodate as much as possible ...

Many architects who were interviewed at the University of Kent described the regulations as generic, vague and ambiguous as they do not explain clearly why such provisions are used and which types of disability such provisions cater for.

> Part M is very generic and it does not say or explain why certain things are done like that. It will be really useful to architects to understand why having, for instance, handrails at whatever height and why certain materials are used.

These generic regulations, as many interviewed architects noted, encourage architects and designers to interpret these standards differently and adhere to minimum standards of provisions. Such standards do not push architects and designers to achieve an inclusive environment (Goldsmith 1997, 2001; Imrie and Hall 2001). Although the regulation standards prescribed in the 2000 Part M Approved Document extend its provisions to cater for different disability types, it follows the top-down approach. As Goldsmith (2001) and Imrie and Hall (2001) state, this approach is centred on accommodating the needs of people with mobility, visual and hearing impairments and ignores the needs of a wide spectrum of users, thus leading to a misrecognition of inclusive design.

Feedback from participants with visual and cognitive impairments and mental health difficulties at the University of Kent stressed the need for prominent visual and audible signage to help in orientation and finding the way around a building, for which Building Regulations, such as Approved Document M, do not provide guidance. This finding is supported by those reported in the literature. Goldsmith (2001, p. 4) points out that paying exclusive attention to the needs of individuals with certain disabilities has hampered the realisation of inclusive design as it has ignored the wide range of users who can be impacted by the same barriers. A focus on compliance with Approved Document M in order to meet design requirements at the University of Kent has led to the exclusion of many users, including individuals with disabilities whose needs cannot be accommodated by these provisions. Geared to achieving and securing accessible environments for a limited range of disabilities (Imrie and Hall 2001), Part M has failed to legislate for the highest possible quality of accessibility to the Universities of Kent, Essex and Bath.

Another limitation of Part M that is supported by Imrie and Hall's (2001) findings is the lack of provision in relation to material finishes, interior decoration and lighting types, which has generated health and safety problems for many participants at the University of Kent. Those with mobility and visual impairments found difficulty in managing slippery and highly reflective floor finishes. On the other hand, the carpet floor finishes tended to cause asthma and health problems for participants with respiratory conditions and were difficult to manage by most of the wheelchair users who were interviewed. By not including design guidance in Part M with regard to decoration, material finishes and signage, many individuals with visual, cognitive and other impairments are excluded from the accessibility enhancement process.

Whilst the Planning and Compulsory Purchase Act 2004 places duties on architects and planners to provide access statements as part of any planning application in order to maximise accessibility, usability and inclusive design, findings from an analysis of the access statement concerning the Jarman Building at the University of Kent revealed that the statement failed to achieve inclusiveness because many of its recommendations were not implemented during the construction and execution phases.

One of the barriers to achieving an inclusive university environment was revealed by the views of architects at the University of Kent. The conflicting requirements of

Approved Document M and British Standards prevented many of them from achieving inclusiveness.

> As a result of a profound observation of all aspects of legislation, I noticed that Part M carries within it many confusing and repetitive information. I think the standards need to be re-quoted. It makes our life so difficult looking at legislation and saying, "Which one is the one? Which one we need to use!!! They have different measurements. Which one takes a precise measurement?"

By only responding to the needs of individuals with disabilities, but without including provisions for other users (Goldsmith 2001), Approved Document M has limited architects in terms of achieving inclusive design at the University of Kent. One such architect who was interviewed gave an example of a lift-manufacturing company that provides inclusive products to accommodate the needs of people with limited dexterity, both seated and standing users; however, she was prevented from using this product as it did not comply with Approved Document M. This finding is supported by those found in the literature (Goldsmith 2001; Imrie and Hall 2001; Steinfeld 1994; Steinfeld and Maisel 2012; Nussbaumer 2012) that reflect the narrow interpretation of inclusive design in the Building Regulations, with the design solutions that facilitate social inclusion, and recognise diversity and difference among potential users being overlooked.

In embracing the principles of social justice, equity, flexibility and choice an inclusive design of a university built environment goes beyond the concept of accessible design. It addresses the issues of difference in terms of users' age, physical and cognitive abilities and diverse cultural backgrounds. By providing creative products and environments that take account of the evolving lifestyles of a wide range of users, their health and well-being, safety, security and social participation, universities could achieve inclusiveness. A good example is illustrated in the strategic design proposal for the Templeman Library. The proposed remodelling of the central atrium has a gently sloped gradient at the main and back entrances offering the user the choice of using either the main or the back entrance to reach the Library. For users who prefer to use the stairs, they have the choice of three different staircases that can accommodate their diverse needs and individual differences. Moreover, the Templeman Library strategic design proposal, illustrated in Chap. 6, offers users the choice of using four different, spacious passenger lifts that are designed to accommodate students and staff members, short and tall people, people with visual and hearing impairments, as well as wheelchair users and mothers with pushchairs or students carrying luggage. In addition, the colour coding system and the information and directional signage proposed at the Templeman Library are all good examples of inclusive design that can benefit all users without patronising the individuals with disabilities.

Rethinking accessibility, as referred to in the Part M Building Regulation and Approved Document M in the UK, by expanding the provisions to cater for children, women, people of short and tall stature, left- and right-handed individuals, people with cognitive impairments and those with asthma and so forth, is another method that could assist architects in understanding the goal of inclusive design and encouraging its adoption.

In the same way that the Equality Act 2010 expanded the requirement to protect people from discrimination in one piece of inclusive legislation, an inclusive Building Regulation could cover the needs of people of different sexes, disabilities, religions, family structure and age groups. By defining the above-mentioned users' needs in the legislation and expanding the building standards to accommodate these needs, the university built environment would no longer be able to discriminate against the categories of users noted above. An inclusive Building Regulation should avoid any conflicting provisions that can create confusion for architects. It should include detailed and clearer information about provisions so that architects will understand the reason for using such provisions for particular users. Moreover, the inclusive regulation should be provided in different formats, such as Braille, large fonts and easy-to-read formats, in order to enable architects, access consultants, access groups and designers to take advantage of this valuable information.

7.1.4 Procedural Barriers

Procedural barriers are another main limitation to achieving an inclusive university environment. The lack of financial resources and knowledge about inclusive design and disability needs have played a major role in the failure to embrace inclusive design.

The main barrier that architects at the University of Kent put forward was economic factors. They believed that enhancing accessibility would add extra costs. Because of the lack of financial resources, they avoided using inclusive provision and technological devices designed to eliminate such barriers. This notion that inclusive design adds cost is a misconception held by many architects, planners and contractors. Such an argument has been challenged by findings from the literature that reveal that buildings designed inclusively from the beginning will need fewer alterations in the future (Levine (2003, p. 10), leading to lower costs later on (Nussbaumer 2012, p. 44). Sawyer and Bright (2007, p. 2) note, 'careful consideration of accessibility issues at the design stage and good management throughout the life of a building can offer and sustain accessible environments at little or no extra cost'. Similarly, maintenance and refurbishment plans present opportunities to resolve accessibility barriers and achieve inclusive environments with minimal or no additional expenditure.

Findings from the study reveal that studying the demographic factors that relate to university buildings confirms that using inclusive design can be a strategy that is good for business. The idea is to inspire and motivate, demonstrating how industrial and commercial enterprises can benefit from a user-centred approach to design and development. Inclusive design could help universities define and implement more meaningful ways in which they could engage with people. This would promote deeper understanding of a particular market sector and what universities might demand from this product or service.

Inclusive design presents many potential benefits for university businesses and architects. For university businesses, expanding the facilities, services and environments and considering the different users' needs increase the university appeal and raise competitiveness. By adopting an inclusive design process, universities would simply be able to comply with current and future legislation (Steinfeld and Maisel 2012; Imrie and Hall 2001). For architects, it could be an opportunity for innovation which could enable them to become pioneers in the inclusive design field. It could allow them to identify problems or issues that have not previously been addressed. Architecture and businesses would have to be sensitive to more diverse markets if they wanted to maintain or expand their appeal. Emerging markets are beginning to become powerful factors in users' attempts to achieve civil and legal rights.

With the growth in numbers of international students at universities, as well as the increase in the number of people with cognitive impairments, universities need to be aware not only of the physical disabilities but also of other impairments and the needs of different users when designing new services or environments.

Inclusive design can help to overcome prejudices against, and traditional views of, target groups and to expand the universities' understanding of the psychology of users. As Steinfeld and Maisel (2012) affirm, improvements in medicine in the current century are leading to people living longer, healthier lives, with a greater burden being placed on social models of healthcare. Moreover, rapid urbanisation and development are creating large cities that struggle with congestion and present an increasing demand for inclusive services and supplies. Furthermore, family structures and changes in them, with members being spread over a greater number of places and countries, are playing a major role in creating business opportunities to satisfy these new demands. To accommodate all these changes, university businesses will need to understand the changing lifestyles of their users and the different contexts in which they are now operating. Aligning business practice with changes in law and policy means that inclusive universities will become the preferred option in the future. To achieve inclusive design, universities should implement inclusiveness not only in their built environments, but also with respect to their learning services. Creating a truly inclusive university differs from the current general discourse on inclusive design as it necessitates greater attention being paid to the needs and activities of a diverse group of users. These include means of transportation used, access to services and facilities, access to knowledge and information, and social integration.

Inclusiveness at a university can be demonstrated by the means of transportation that is available, parking bays, external routes and staircases, and external seating areas. It should also be reflected in the interior design of buildings including provision of adjustable furniture, hearing aid systems in all interior facilities, soft finishing materials with good visual contrast and material finishes that do not cause allergies or health problems, in addition to toilet compartments for individuals with disabilities, families, children and females and males. Universities could achieve inclusiveness inside buildings by adopting inclusive signage systems, with text, pictograms and Braille characters on internal doors, directional and information signage, and emergency exit routes. Moreover, providing emergency exit systems and passenger lifts with audiovisual announcements and buttons with Braille and embossed characters could benefit all potential users when navigating independently, both inside and outside buildings. Inclusivity could also be observed within the learning services by providing alternative text formats, including large print, Braille, audio tape, easy-read materials and electronic files. Table 7.2 illustrates proposed inclusive design criteria that universities should adopt in order to respond to their users' needs.

A major barrier facing architects at universities is the lack of knowledge regarding the needs of people with disabilities. Many of them have never attended training courses about individuals with such needs. To tackle this lack of knowledge, universities should include courses in their educational programmes that are mandatory for all design and architect students. It would also be highly desirable for the governing body or board to which professional architects belong to give incentives to architects to attend seminars on inclusive design and disability awareness, for example, by reducing the board registration fees or taxes. These seminars would be more effective if run and delivered by a wide range of users with disabilities. Involving a variety of individuals with disabilities in the delivery of training courses and allowing them to share their experiences would break down prejudice and sociocultural attitudes and would help professionals to enhance the built environment and the services provided (Lifchez 1987; Imrie and Hall 2001).

7.1.5 Organisational Barriers

Inclusive design policies at universities are greatly lacking. Educational organisations do not acknowledge the users' needs nor do they give incentives for the inclusive design approach to be adopted. Their understanding of inclusive design is centred entirely on eliminating physical barriers and enhancing the level of accessibility for wheelchair users, disregarding the broader range of other users' needs.

Findings reveal that physical barriers may be relatively easy to overcome, but without an inclusive management policy and practice, universities will fail to achieve inclusiveness. By providing appropriate and effective management training courses for all staff members and employees, universities would be able to provide inclusive environments, transportation, services and information that are inclusively managed and maintained. The training should cover working with individuals with disabilities, legislation, inclusive building regulations, disability awareness and how to make written information accessible and understandable for individuals with disabilities. Universities must establish procedures to ensure that when opportunities to improve access arise, appropriate management procedures and practices are implemented. Identifying where maintenance or management input is required is crucial to making and sustaining inclusive environments (Sawyer and Bright 2007, p. 39).

Table 7.2 Proposed inclusive design criteria

Site topography: Site topography is an essential element of inclusive design that aims to reduce the effort needed to access buildings and facilities
• The site should be graded to eliminate steps and stairs.
• Main entrances should be relocated in order to be more accessible and easy to manage.
• Proposed accessibility provisions should not reduce the historical value of a site or building.
External pedestrian routes and pavements
• They should be easy to navigate and provide access to all facilities.
• If there is more than one route to a facility, both the primary and the secondary routes should be easy to access and clearly signposted.
• Pedestrian crossings should be clearly marked and signposted.
• Pedestrian, vehicular and bicycle routes should be separate. If they are adjacent to one another, they must be clearly marked and signed.
• Steps and kerbs should be avoided along pedestrian and bicycle routes as they can be a hazard for wheelchair users, mothers with pushchairs and people with visual impairments.
• Seating and rest stops along routes should be away from the pedestrians and the routes they use.
• Pedestrian routes must have firm, slip-resistant and low-reflective surfaces.
• There should be tactile blister warnings on dropped kerbs to assist individuals with visual impairments and wheelchair users.
• Where a bus drop-off zone is provided along a pedestrian route, the bus should provide level access to the pedestrian route, and it is recommended that a bus stop with a covered canopy and seating should be provided.
• The bus stop should be clearly signposted with text fonts that contrast with the surroundings.
• Pedestrian routes should be graded to facilitate the movement of wheelchair users. Major access routes should not have slopes whose gradients exceed 5%. Any steeper gradient must be designed as a ramp.
• If routes are steep and have difficult surfaces or barriers, information should be provided for users who are not familiar with the route. Alternative and shorter direct routes should be recommended so users can follow those instead.
• Pavements should be made wide enough to accommodate two wheelchairs, and if a bus stop is provided, further widening of the pedestrian route should be considered.
• Routes should be free of obstructions and hazards. Drainage, signs, overhanging trees, light fixtures or benches should be eliminated.
• Audible beacons or talking signs should be installed to aid individuals with visual impairments.
Street crossing to accessible routes of travel
• Kerb cuts and ramps should direct pedestrians into safe crossing areas.
• Pedestrian crossing distance should be reduced on major crossing routes by providing 'bumps' at the corner or safe islands between central lanes. This reduces exposure to traffic, provides a safe place to wait and helps to minimise hazards.
• Crossings should be clearly signposted with signs that have adequate visual contrast and text fonts.
• Pedestrian crossing signals should provide enough time for people who move slowly to cross. They must be clearly detectable by motorists.

(continued)

Table 7.2 (continued)

• Visual and audible pedestrian crossing signals should be provided. Lower pitched signals are favoured over high-pitched ones as they can be easier to hear.
• Higher illumination must be provided at pedestrian crossings.

Car parking and buses

• The number of accessible parking bays should match the number of potential users.
• Accessible parking bays should be placed close to main entrances.
• Accessible parking bays should be clearly signposted and marked with hatched zones for rear access.
• Accessible bays should have level routes to main entrances of buildings.
• Drop-off car zones for cars should be provided close to main entrances.
• Pathways should be well lit and provide protection from the weather.
• Directional and informational signage should be provided near parking bays in visual and tactile forms.
• Maps with information about accessible parking bays and routes to buildings should be provided with visual, tactile and audiovisual forms.
• Family, visitor and motorcyclist bays should be provided near the main entrance to buildings.
• Tactile maps should be provided close to car parking and bus stops.
• Buses and bus shuttles should be inclusively designed to accommodate the needs of all users.
• Level access should be provided in all buses.
• A dedicated space for wheelchair users and mothers with children and pushchairs, as well as spaces for storage and luggage, should be provided in buses.
• Adequate visual contrast should be provided between bus-floor surfaces and walls.
• Audiovisual announcements and call buttons should be provided in buses to alert people with visual and hearing impairments about impending arrival at the destination.
• Slip-resistant floor surfaces should be provided for bus-floor finishes.

External ramps and stairs

• Ramps or gentle slopes should be used as the main circulation route, with adequate width to accommodate pedestrians and wheelchair users.
• The ramp should run along the direction of travel to reduce the need for extra effort.
• Gentle slopes should be provided so all users can manage easily without extra effort.
• Ramps should be wide enough to accommodate the needs of all users.
• Ramps should be designed so they can be used by all people rather than just wheelchair users.
• Handrails should run along the ramp and landing and should be placed at two heights (900–1000 mm from floor level) for adults and (500–600 mm) for children and seated users.
• Handrails should have a surface material that is easy to grip and contrasts well with the background surfaces.
• If ramps are long, a level resting landing should be provided.
• Stairs should not have more than ten risers on stairways between landings.
• The use of tampered, curved steps should be avoided.
• Stairs with open risers should not be used as they can result in trip hazards for users.
• External stairs should have tactile corduroy warnings on the top and bottom of steps to assist individuals with visual impairments.

(continued)

Table 7.2 (continued)

• Consistent risers and treads should be provided to reduce trip hazards or the risk of falling, and to support a comfortable pace.
• Risers should have clear and contrasting nosings at the top and edge of the steps.
• Step surfaces should be made from slip-resistant material but without excessive friction.
• Stair landings should be wide enough to provide extra space for resting.
• Lighting conditions should eliminate strong shadows on stair treads at night.
• Photoluminescent stripes should be provided to assist people with visual impairments in identifying stair nosings and treads at night.
• Handrails should run along steps and landings, extending 300 mm at the top and bottom.
• Handrails of steps should be placed at two heights, as recommended for ramps.
• Handrails should have a surface material that is easy to grip and contrasts well with the surrounding background surfaces.
• Ramps and stairs should be marked and well lit.
Cash machines and telephones
• Cash machines such as ATMs, ITMs and university public Internet access points should be within the reach of wheelchair users, and people with short stature. They should provide audible announcements for people with visual impairments and have easily operable controls for people with cognitive impairments. Payment methods should be provided in the different systems, which are adaptable to seated and standing users, and those with different sensory abilities and different languages.
• Telephones provided at a university used by the public should be compatible with hearing aids, have volume control and be placed within reachable height for wheelchair users.
• Public telephones should have shelves to assist users who wish to take notes. It is recommended that an adjustable seating area should be provided.
Main entrances
• Main entrances should be clearly signposted and landmarked, with adequate lighting.
• Motion-detector-controlled lighting along the route to the entrance should be provided.
• Entrances should be positioned to provide a short and direct route to the main destinations in the building.
• They should have canopies over them to facilitate entry and exit in adverse weather conditions.
• If main entrances have glazed automatic doors, they should have clearly contrasting visibility enhancements at two heights so they are visible to all users.
• If a building has two main entrances, it is recommended that each entrance door should be clearly identified through the use of different graphics, for example employing a number or colour coding systems.
• Information and directional signs should adopt a colour coding system and should also be provided in Braille and embossed characters together with audiovisual signage that is placed in main entrance lobbies and internal routes.
• Information about entrance/exit locations should be available on the university website, including routes undergoing construction and alternative routes.
Lobby area
• A waiting and social interaction area should be provided in the lobby area.
• Furniture should be flexible and easy to manage and maintain, and should accommodate a wide range of needs (chairs with/without armrests, low chairs, high chairs, adjustable tables, movable tables, fixed tables, etc.).

(continued)

Table 7.2 (continued)

- The layout should be planned so that it accommodates several different furniture configurations.

- Space for wheelchairs should be integrated into the seating arrangement.

- If a screen is provided, the distance between the seating area and LCD screen should be three times the size of the screen.

- If water fountains are provided in lobbies or corridors, it is recommended that they should be adjustable to alternative heights for seated and standing users.

Reception area

- It should be located in a place where external noise is reduced but it is still visible to visitors.

- It should be clearly signposted and positioned so the receptionist and visitors can clearly see each other.

- It should be positioned away from windows or glazed doors to avoid creating confusing shadows and glare for people who rely on lip-reading or sign language to communicate.

- Reception counters should be set at a height suitable for both seated and standing users, with a low section (700 mm from floor level) and sufficient counter space to allow people to write or sign documents.

- The low counter should be clearly visible from the main entrance and accessible from both staff and visitor sides.

- The counter worktop should contrast with its edge and the surrounding surfaces, and its exposed edges and corners should be well rounded.

- Induction loops and speech enhancements should be provided in the reception area to serve staff and visitors. The international signage for induction loops should be provided and fitted on the reception desk.

Internal circulation

Corridors, signage, vertical circulation

- Corridors should be wide enough to accommodate two wheelchairs and people walking in different directions.

- Handrails should run continuously along corridors without interruption, except at doorways and openings, and should have recessed brackets so that a hand can move from end to end without interruption. These can then serve as a means of navigation for people with visual impairments.

- Removal of overhanging hazards should be facilitated by the provision of recessed openings on corridor walls to create pathways free from obstruction.

- Columns along passageways should be made clearly visible when seen against background surfaces by incorporating a band of contrasting colour or luminance at two heights (1500–1650 mm) and (850–1000 mm).

- Wayfinding systems and signs should be provided for first-time visitors to guide them along corridors.

- Information that can be easily updated should be provided.

- A signage system should be selected that provides enough flexibility to allow it to be used in different types of building. Pictograms and words in Braille and embossed text provide such flexibility and diversity.

- Lights should be carefully studied so they do not create reflections and glare on signage.

- Lighting effects should be used to highlight paths and distinguish them from one another.

- Colour coding, material finishes and textures and patterns should be used to provide sensory cues for all users to help them distinguish the facilities in interior settings.

(continued)

Table 7.2 (continued)

- The vertical circulation should be considered whenever renovating existing buildings. Adding vertical circulation to an existing building could tackle an accessibility barrier and offer alternatives for potential users.

- It is recommended that internal staircases should have the same provisions as external staircases, but without tactile corduroy warnings.

- Passenger lifts should be clearly signposted from the main entrance. The term 'disabled lifts' should be avoided as lifts should be usable by all people rather than just a specific group of people.

- Lifts should have landings large enough for wheelchair users to turn and reverse into a lift.

- The floor area outside the lift and inside it should be visually distinguishable from surrounding wall surfaces.

- Lifts should have an automatic door wide enough to accommodate wheelchair users and standing people.

- The 'lift coming' indication should be positioned at a height that is clearly visible to all users.

- It is preferable for control buttons to be located on both sides of lift walls, placed at a reachable height for seated and standing users.

- Call and control buttons should have embossed and Braille characters or numbers and should visually contrast with the lift's background surfaces.

- Audible announcements and visual displays should be available both internally and externally to indicate which floor the lift has reached and the facilities on each floor or to inform users that the doors are open.

- Emergency intercom telephones with inductive couplers should be installed in the lift.

- Alarm buttons in the lift should be fitted with a visual acknowledgement and flashing beacons to indicate to lift users unable to hear it that the alarm has sounded.

- The lift should have a slip-resistant floor surface and wall surfaces that minimise glare and reflections.

- Adequate lighting should be considered so it does not cause glare or create confusing shadows and pools of light and dark on control buttons and inside the lift.

- Lifts used for emergency evacuation should be clearly signposted and fitted with an independent power supply.

Toilet compartments

- Toilets should be clearly signposted and have signs with embossed letters and Braille characters, placed on one side of the toilet doors, at 1000 mm from floor level.

- The toilet door frame and panel/leaf should be clearly visible using contrasting colours.

- Doors should be easy to open and have lever handles that are clearly distinguishable from surrounding finishes and placed at 1000 mm from floor level.

- Cubicles should be wider than 800 mm and have grab rails.

- Cubicle doors should open outwards to provide more space.

- Sanitary units should be placed at reachable heights, namely 600–800 mm from floor level.

- Adjustable grab rails or alternative configurations should be provided to accommodate different needs.

- Extra space to accommodate students' books, briefcases and laptops should be provided.

- Shelves should be provided to be used as temporary storage for personal items.

(continued)

Table 7.2 (continued)

• Adjustable WC seats should be provided, placed at different heights and with transfer directions to accommodate the needs of different users, namely right-hand and left-hand transfer.
• Basins with pipe systems should be mounted on walls to provide adequate manoeuvring space for wheelchairs.
• There should be lever taps with identification signs.
• Easy-to-manage fixtures and sanitary units should be placed at reachable heights and should contrast well with the surrounding backgrounds.
• There should be unisex accessible toilets with a large space to provide room for companions and for large wheelchairs and devices to manoeuvre.
• Automatic lighting should be enhanced to ensure that there is adequate visual contrast between the sanitary units, fixtures, walls and floor finishes.
• The emergency alarm system should be easy to manage and reach.
• Family toilet compartments for both adults and young children should be provided and clearly signposted. The compartment should include a nursing room for mothers.
• Toilet compartments should include a washing and showering area.
Emergency exit routes
• Consistent emergency exit signage should be provided within every campus building and along emergency exit routes.
• Audible announcements and visual displays should be placed internally and externally on the exit routes to alert users with visual and hearing impairments.
• Fire hoses should be placed in recessed storage areas along the passageway to avoid obstruction.
• Refuge areas should be clearly signposted and free from obstruction.
• Emergency exit staircases should be clearly signposted and have clearly marked nosings on steps and handrails running along the top and bottom of steps.
• Adequate lighting should be installed and checked to see that it does not cause glare or create reflections.
• Photoluminescent stripes should be provided to assist in identifying stair nosings and treads at night.
• Level exit routes should be provided.
• Emergency evacuation lifts in libraries and crowded facilities should be provided. Such lifts should be clearly signposted.
Inclusive management practices
University buildings contain different types of facility, including libraries, refreshment areas and restaurants, sports centres, museums and exhibition rooms, residential accommodation, lecture theatres and auditoriums, which attract a very diverse set of users and visitors. To satisfy their diverse needs in an inclusive university campus, a set of design strategies and management policies is suggested as follows:
• The university website should be inclusively designed to include a section which provides information for all visitors about the environmental barriers they might encounter on site and the features available to support their visit. It is recommended that the university website should include the facilities provided on its site plan, such as bus stops, accessible car parking spaces, car drop-off zones, main entrances, level access, ramps and staircases.
• The university should adhere to the comprehensive guidelines to ensure that its website is accessible to all users.

(continued)

Table 7.2 (continued)

• Information handouts, leaflets and brochures in multiple languages and different formats and text font sizes should be available.
• Staff members and education providers should receive training about different disability needs, how to work effectively with individuals with disabilities, recruitment policies and how to attract individuals with disabilities.
• Training courses should also include sessions about up-to-date legislation and codes of practice.
• It is recommended that architects collaborating with the Estates Department should attend training courses dealing with disability awareness, Building Regulations and working effectively with access groups and individuals with disabilities.
• Training courses should also be delivered by individuals with disabilities to raise awareness.
• Departments within the university should collaborate and exchange information about disability needs and ways to enhance accessibility and promote inclusion. Collaboration between the Estates Department and Diversity Office, Disability and Dyslexia Student support and the student union, for instance, could take account of different perspectives and approaches when promoting inclusive environments not only by enhancing the physical environment, but also by working on providing an inclusive educational curriculum that accommodates the needs of all users according to their abilities.
• It is recommended that a university should carry out online surveys to obtain feedback on students' and staff members' experiences of accessing the services and facilities provided. This would include receiving feedback about the physical environment and barriers encountered in management practices and procedures, disability awareness among staff members and lecturers, accessing information and the website, etc.
• Consultation with individuals with disabilities is recommended as the starting point whenever a university is going through a site development. It is recommended that consultation should be arranged between the Estates Department and policy-makers at the university. Involving students and staff members familiar with the site would assist professionals and architects in working together to create an inclusive environment. This consultation should start at the early stages of pre-planning, selecting materials and developing concepts, and continue until the last stages of construction.
• Introducing inclusive design competitions in university architecture, art and design, and engineering departments is a key element in raising awareness about the inclusive design approach and ways to implement its principles and there should be collaboration with disability offices at the university, such as the Disability and Dyslexia Office at the University of Kent. Involving students in such competitions and working with a broad range of students with disabilities at the university on dedicated projects can lead to achieving inclusive designs.

In addition, another barrier to achieving inclusiveness at universities is the failure to broaden and provide knowledge and information in an educational setting. This barrier must be overcome. As an example, instructors will often fail to recognise the needs of a range of users, including those with visual and cognitive impairments, and will overlook the need to change their teaching methods or provide large-print materials or easy-to-read documents.

In order to achieve an inclusive environment, universities should implement an inclusive and holistic approach throughout their educational function and their environment. Inclusive design cannot be achieved in an educational organisation without meeting the requirements of inclusive education. The concept of inclusion in both the physical and educational environment must be recognised and the users' and learners' needs must be accommodated in order to maximise social justice, democracy and social participation. This can be achieved by changing instructors' attitudes and raising awareness of diverse disability needs. Inclusive technology is the main driver in enhancing the learning environment at universities. Through training and provision of modified keyboards, speech enhancement instruments, text-to-speech, notes with large print and easy-to-read notes, in addition to Ipad tablets, software and other 'high-tech' gadgets and devices, instructors can change and adapt to these demands in order to accommodate the diverse needs of their students. Establishing a centre for inclusive education is crucial to maintaining and providing inclusive services for all users, including people with disabilities.

An inclusive design strategy does not need to be limited to the design process. It can be a foundation on which universities can base their entire business philosophy. Not only should a truly inclusive university be reflected in its built environment, but it should also be seen in the employment practices, services and information that are accessible to potential users.

An inclusive way of thinking can influence not only the physical features of buildings, but also employment policy, personnel management, customer service, educational curriculum, communications strategy and marketing. This implies that universities can promote and adopt inclusive design throughout their departments and services. By formulating a holistic approach that roots itself in university culture, policy and practice, inclusive training courses could be delivered to each department, which would enable them to target their facilities to accommodate the needs of all potential users, including those with disabilities. Moreover, introducing inclusive design into curricula in the architecture, engineering, and interior and graphic design departments, plus involving students in proposing inclusive design solutions for existing and new buildings, would enable students to take part in enhancing and shaping their university built environment. Such a recommendation was suggested by users with disabilities and architects who were interviewed. The architects pointed out that the paucity of knowledge about accessibility was one of the reasons environments have been created that do not acknowledge the different users' needs. Introducing inclusive design into the educational curriculum would lead to a better understanding of inclusive design among students and would contribute towards developing inclusive solutions that can offer cost-effective and uncomplicated methods of achieving innovation. This finding is supported by Steinfeld and Maisel (2012, p.74) who stress the importance of integrating inclusive/universal design in professional educational programmes to increase the rate of the adoption of inclusive/universal design.

Another main contributor to mainstreaming inclusive design in an educational setting would be to introduce the extent of the adoption of its principles in university ranking criteria. Inclusive design principles could be part of the ranking criteria when inclusive policies, practices and services were audited. Universities adopting these inclusive design criteria in their built environment and educational system would be given more points in their ranking. By fulfilling this requirement, inclusive design would become mainstream in an educational setting.

References

Burton, E., & Mitchell, L. (2007). *Inclusive urban design: Streets for life*. Oxford: Architectural Press.

Finkelstein, V. (2002). The social model of disability repossessed. *Coalition*, February, 10–16.

Goldsmith, S. (1997). *Designing for the disabled, the new paradigm*. Oxford: Architectural Press.

Goldsmith, S. (2001). *Universal Design. A manual of practical guidance for architects*. Oxford: Architectural Press.

Gooding, C. (1996). *Blackstone's guide to the Disability Discrimination Act 1995*. London: Blackstone Press.

Imrie, R., & Hall, P. (2001). *Inclusive design: Designing and developing accessible environments*. London: Spon Press.

Levine, D. (2003). *Universal design New York 2*. New York: Center for Inclusive Design and Environmental Access. University at Buffalo, State University of New York. Retrieved July 30, 2018, from www.nyc.gov/html/ddc/downloads/pdf/udny/udny2.pdf

Lifchez, R. (1987). *Rethinking architecture: Design students and physically disabled people*. London: University of California Press.

Nussbaumer, L. (2012). *Inclusive design: A universal need*. New York: Fairchild. London: Bloomsbury.

Planning and Compulsory Purchase Act. (2004). Retrieved July 30, 2018, from https://www.legislation.gov.uk/ukpga/2004/5/section/29

Steinfeld, E. (1994). *The concept of universal design*. Retrieved July 30, 2018, from http://www.udeworld.com/dissemination/publications/56-reprints-short-articles-and-papers/110-the-concept-of-universal-design.html

Steinfeld, E., & Maisel, J. (2012). *Universal design: Creating inclusive environments*. Hoboken, NJ: John Wiley & Sons.

Swain, J., & French, S. (2000). Towards an affirmation model of disability. *Disability & Society, 15*(4), 569–582.

Sawyer, A., & Bright, K. (2007). *The Access manual: auditing and managing inclusive built environments*. Oxford: Blackwell Publishing Inc.

Woodhams, C., & Corby, S. (2003). Defining disability in theory and practice: A critique of the British Disability Discrimination Act 1995. *Journal of Social Policy, 32*(2), 159–178.

Chapter 8
University Inclusive Environment as a Future Vision

8.1 Introduction

An inclusive university is a place which integrates all its services and facilities so all users, including individuals with disabilities, can reach its buildings, access the same doors and use the same routes and facilities. It is a place where users do not feel they are on show when using services. It is a place where one can reach and use the facilities independently without worrying about parking spaces, orientation and wayfinding, entrances, toilet compartments and emergency exit routes. It is a place where I will be proud and really happy to say it doesn't matter what aid I am using, whether crutches, manual wheelchair, with a companion or an electric wheelchair, canes or guide dogs, the university is fully accessible wherever you go and whenever you want to. That is my ideal scenario (Student at the University of Kent, June 2009).

The student's observation and ideal scenario affirm the conclusion of this book, namely that universities are still not fully practising inclusive design. Whilst efforts have been made at the University of Kent to eliminate physical barriers and improve the level of accessibility, it has failed to provide completely inclusive environments and services. This chapter presents general conclusions, recommendations and guidelines stemming from the qualitative and quantitative methods used. These recommendations and guidelines, if put into practice, could contribute towards creating inclusive universities. This chapter aims to explain how inclusive principles can be put into practice to achieve a friendly and welcoming university environment.

8.2 Further Work

The vision of an inclusive university is reflected in the thoughts and opinions of all users, namely students, staff members, stakeholders and architects, collected during the current study. All these reflections are used to make recommendations to achieve that goal; however, the research has clear limitations.

© Springer Nature Switzerland AG 2020
I. Shuayb, *Inclusive University Built Environments*,
https://doi.org/10.1007/978-3-030-35861-7_8

- At different times during the project the researcher was able to reside in close proximity to the University of Kent. This factor, allied with a limited time frame and research funds, was the rationale for conducting the research on this case study and not extending it to other universities. The researcher carried out in-depth access audits on six building types at the university, but, nevertheless, to feel truly confident about proposing inclusive recommendations, further research should be carried out on a larger number of universities with different topographical sites, building types and period of original construction. Through such research, one could measure and record the design characteristics of each university to make more precise proposals about how to achieve an inclusive university environment, which could be used as guidance for architects and professionals.
- The recommendations from this study are based on input from a sample of 10 individuals with disabilities, 6 stakeholders and 174 online respondents with and without disabilities at the University of Kent. The researcher interviewed these people in depth, explored the physical and management barriers they encountered, walked with many of them to newly erected buildings and fed them the findings from the access audits. However, talking to many more people with different needs, ages, abilities and cultural and ethnic backgrounds would assist in proposing inclusive recommendations that could tackle all these diversities.
- The research focused on the accessibility of the external and internal environments in campus in order to give an idea of the factors that limit accessibility, but it did not focus on other important aspects. Further research could be carried out, focusing on outdoor environments, such as roads and pavements, and outdoor spaces, like gardens, across the campus, to determine how users interact with the outdoor environment and how such interaction can promote inclusiveness.
- The inclusive university recommendations are centred on general aspects of enhancing accessibility in the campus, such as means of transportation, pedestrian routes, car parking spaces, internal facilities and emergency exit routes. However, they do not cover the neighbourhood surroundings, such as streets, public spaces, schools, leisure attractions and other facilities. Further research should study how these aspects can interact to achieve an inclusive environment.

8.3   Recommendations for Achieving Inclusive University Environments

The research reveals that the exclusive design approach still takes precedence over the inclusive one in universities. Four barriers have been identified that hinder the adoption of inclusive design. The sociocultural differences, the failure to define inclusive design and disability, the accessible design and regulation barriers, as well as the procedural and organisational barriers have all led to the creation of exclusive environments at the University of Kent.

To eliminate these barriers, this research calls for a new paradigm, illustrated by diverse strategic inclusive design case studies and access audit documentation, which could be used as good examples of how to achieve inclusive university environments. Moreover, the research proposes five key recommendations that could lead towards a better understanding and implementation of inclusive built environments at universities. These recommendations are (1) rethinking disability and inclusive design, (2) revising regulations, (3) developing distinctive procedural plans for new and existing buildings, (4) filling the gaps in the process of mainstreaming inclusive design and (5) establishing criteria for the achievement of an inclusive university built environment.

8.3.1   Rethinking Disability and Inclusive Design

To achieve an inclusive university environment, the sociocultural as well as economic and organisational attitudes towards disability should be understood in terms of the users' civil, moral and legal rights. The definition of disability should be understood in terms of human differences and their interaction with sociocultural and economic and organisational attitudes and barriers. Rather than defining disability as a reality that sees the individual's personal problem in terms of how to adapt to the existing built environment, it should be understood in relation to the people's different qualities and their need to integrate into society. Hence disability should be comprehended as what is wrong *for* the person and not what is wrong *with* the person.

A better understanding of inclusive design and disability in terms of accommodating the diversity of the human life cycle, lifestyle and culture will lead to design spaces that accommodate these variables. Hence, the proposed new definition of inclusive design is centred on these variables rather than merely acknowledging the needs of individuals with disabilities and elderly people. The researcher proposes the following inclusive design definition:

> Inclusive design is a process that takes account of the diverse human, cultural and life-cycle factors to improve human interaction, social participation, health and wellbeing, by involving the user in the design and development process.

The proposed new definition is an improvement on existing definitions, as it includes the main principle of inclusive design and the goals of universal design that Steinfeld and Maisel (2012) propose. By acknowledging human diversity, cultural lifestyles and abilities and by involving potential users in the design and implementation phases, environments and products can be adapted to the users' needs, will be safe and healthy to use, and will be convenient for all potential users.

8.3.2   Revising Regulations

The research has described an important aspect of inclusive universities, which is often overlooked in the literature on this subject. Whilst the concept of inclusion is becoming more vital with respect to enhancing accessibility and its principles are being applied in public spaces, this research has shown that Building Regulations with minimum requirements, such as Approved Document M, which have been followed at the University of Kent, have failed to create an inclusive environment. A different approach to designing the university built environment has to be developed and adopted.

The research has proposed that inclusive Building Regulations should include more information about provisions to satisfy users' specific needs, as well as provisions for children, women, mothers with children, etc. The regulations should include information about inclusive means of transportation and their specific dimensions and features, inclusive vehicle parking, external and interior features, acoustics, lighting, signage and inclusive information and communication systems. Such regulations, oriented towards a user-centred approach, would assist architects in designing products, provisions and environments that would reduce the need for specific accommodations and would avoid any stigma.

To achieve an inclusive environment at the university, the Estates Department at the University of Kent should strive for higher standards of, and criteria for, inclusiveness (illustrated in Table 6.1).

These criteria aim at implementing a higher level of accessibility to meet the needs of a wide range of users with different abilities. For example, providing a passenger lift with buttons placed at an adequate height for all users, including those seated or with dexterity limitations, would eliminate the possibility of excluding some users.

The research has proposed revising the current Building Regulations to avoid any conflicting provisions that might create confusion for architects. The revised regulations should include detailed and clearer information about provisions so that architects will understand the reason for using such provisions to help particular users. Moreover, the revised regulations should be provided in different formats, such as those using large fonts and easy-to-read formats, to enable architects, access consultants, access groups and designers to take advantage of this valuable information.

8.3.3   Distinctive Procedural Plan for New and Existing Buildings

The general consensus among the architects interviewed at the University of Kent was that eliminating architectural barriers within existing buildings is more challenging than eliminating barriers in new ones. This research has distinguished between existing and new buildings at the University of Kent when it comes to applying inclusive design and procedures.

The ultimate aim of this research has been to improve the level of access to university buildings for all potential users, including individuals with disabilities. To achieve this goal, inclusive design for university buildings should have several layers of analysis included in the university master site plan. There should be a procedural plan which would be implemented over time at universities. Users should be involved in the development of this action plan in order to prioritise actions and enhance accessibility to most currently inaccessible buildings.

This research has revealed that Building Regulations should be used as a tool to solve accessibility barriers and not as a means of restricting architects with respect to innovations. Universities aiming to create inclusive environments in their campuses are recommended to take any renovation projects for their existing buildings as a golden opportunity to apply the inclusive design principles. To tackle the accessibility barriers at an existing or old building, architects should identify the building's strengths and its limitations by studying its interaction with the surrounding built environment and how it is used and managed by its users. Another important element that leads to achieving inclusiveness at existing/old buildings involves acknowledging whether the building is undergoing an extension of its existing structure or whether a new space is being added to it. Extensions and changes to circulation and orientation arrangements in such a building, along with renovations to its older parts, could radically change the way an existing building functions internally and externally and the way it can interact with the surrounding buildings in campus.

Different case studies, all centred on the inclusive design approach, offer the reader new insights into adopting inclusive design at a number of existing buildings, the aim of which was to promote health and safety and social participation without affecting the fabric of these buildings. The procedural plan for an existing 1960s' building, which was going through a process of extension, is illustrated in the case study of the Templeman Library at the University of Kent. It shows how such a building can achieve inclusiveness by adding horizontal circulation in the main central block, thereby resolving the main accessibility barrier between the front and back entrances of the building. The strategic inclusive design proposal not only tackled the physical accessibility barriers, but also created a space that can be easily managed by all potential users. Moreover, the extension phase was used to propose suitable furniture, interior decoration and technological devices that could be adjusted according to individual preferences.

Another example, Eliot College at the University of Kent, is an existing 1960s' building that is used as a work, study, accommodation and social entertainment space. This case study illustrates how the issue of orientation and navigation at an existing building can be achieved by applying inclusive design principles without going through major structural alterations or extensions. Adopting a colour coding system for walls and floor finishes in an existing building can resolve the orientation issues and achieve an easy-to-manage environment that can benefit all users, including people with visual and cognitive impairments. Moreover, the gentle-gradient slope proposed for Level 3 is designed to provide an accessible main entrance link-ing Eliot College with the university campus routes and buildings.

The two strategic design case studies illustrate flexible ways in which inclusive design can be adopted in order to tackle different accessibility barriers in existing University of Kent buildings. Each case study shows that inclusive design is an ongoing process that is not static, but fluid, and is essential when revaluating build-ings after they are occupied.

8.3.4   Filling the Gaps in the Process of Mainstreaming Inclusive Design

Achieving inclusiveness in university building is a continuing process that is affected by the changing demands of users and their needs. A university, to be inclu-sive, has to acknowledge such demands by obtaining feedback from users about their specific needs and providing up-to-date new technologies and designs that cater for such needs and diverse user groups. Potential users must be involved in the design process from pre-planning until the construction phase, as well as in the choice of materials.

Architects aiming to achieve inclusiveness at universities are recommended to use regulations as well as engage their creative imagination by involving stakehold-ers and users rather than focusing on a rule-based approach. They should acknowl-edge that inclusive design broadens accessibility and should not be regarded only as a form of regulatory constraint. Their efforts to design for diversity and individual differences should lead to creative and innovative solutions. They should also assess how their inclusive buildings are functioning after occupation. Such ongoing evalu-ation can lead to achieving a more inclusive university environment.

Universities aiming to achieve inclusive design and comply with the legislation should check that implementation of access audit recommendations is accurately observed. Quality control of the executive team is crucial when it comes to checking compliance. Making accessibility a shared task is vital if inclusive environments are to be achieved. Proper training for the executive team concerning accessibility and inclusive standards is important in the implementation phase. Achieving an inclu-sive environment should be the equal responsibility of access auditors and consul-tants, inclusive designers, architects and executive teams to ensure that inclusive design becomes part of the university's culture.

To respond to the needs of all users and improve the quality of access and levels of enjoyment and satisfaction for every potential user of the university built environ-ment, the research conclusively shows that architects should include students, staff members, librarians, maintenance crews and so forth in the design process.

A golden opportunity to achieve an inclusive university environment has been grasped via the flagship project that has taken place at the Canterbury campus of the University of Kent. This project aimed to add a 4000 m^2 extension to the Templeman

Library, as well as create major enhancements to the existing building. Involving all users, including individuals with disabilities registered at the DDSS, in the design process from the beginning was the first step towards achieving an inclusive library at the Canterbury campus. To promote inclusive design in this project, the researcher contacted the project committee members and head of the Architecture Department at the University of Kent. She also submitted a copy of the proposed inclusive strategies illustrated in Chap. 6 and her recommendations for achieving an inclusive library environment. A positive response to this recommendation was received in that the University of Kent contacted students and staff members to involve them in the consultation. Further information about the project is posted on the University website at http://www.kent.ac.uk/is/projects/templeman/.

Universities basing their entire business on the inclusive design strategy should apply it during both the design and the implementation stages. Moreover, inclusiveness should be recognised in both the physical and educational environments, with the users' and learners' needs being accommodated in order to maximise social justice, democracy and social participation. Establishing a centre at universities for inclusive education and services is crucial to maintaining and providing inclusive services for all users, including people with disabilities. Raising awareness about inclusive design is vital to ensuring that it becomes mainstream at universities, and they are recommended to introduce inclusive design principles into their educational curricula in the architecture, engineering and design departments.

8.3.5   Inclusive Criteria for an Inclusive University Built Environment

University buildings designed from the beginning to be usable by everyone are likely to need fewer renovations in the future and their construction will not cost as much when compared with the expense of enhancing the accessibility of an existing building. Acknowledging that the market for universities is changing radically, and there is a need to attract students and staff members from all over the world, this research proposes that guidance should be based on the principles of inclusive design, which, if adopted, could lead to achieving inclusive universities that could attract more students and staff members to enjoy the diverse and inclusive environments. These criteria could apply when adapting an existing physical environment or when redeveloping, extending, altering, replacing, refurbishing or constructing university buildings, regardless of their date of construction.

The key recommendations relating to physical environments have been noted after consulting with people with different abilities, including people with disabilities. Some of the guidelines have also been taken from Part M regulations as their provisions cater for people with physical, visual and hearing impairments.

8.4   The Importance of Inclusive Design in University Settings

This research has several major strengths and significant ways in which it contributes to the existing literature.

8.4.1   Inclusive Design: A New Paradigm for University Settings in the UK Context

To begin with, this study presents a breakthrough in the fields of inclusive design in educational/university architectural contexts. In the UK, a good example of a developed country, the Equality Act 2010, has put legal and moral obligations on public universities to provide equal opportunities for individuals with disabilities to gain access to employment, education and services, and this research has specifically helped to create inclusive environments at the University of Kent. It took an active role by proposing an inclusive strategy for the Templeman Library, which has undergone an extension. A copy of this proposal and recommendations for the Templeman Library and five other existing buildings (illustrated in Chaps. 5 and 6) was submitted to the Estates Department, Templeman Project committee members and the head of the Architecture Department at the University of Kent. These access audit reports, strategic design proposals and recommendations all contributed towards the creation of environments that are accessible for all.

8.4.2   Inclusive Design Process Versus Inclusive Society at Universities

The ultimate aim of this research was to acknowledge the challenges and barriers that have hindered the achievement of inclusiveness at the University of Kent. The study has contributed towards understanding the sociocultural, economic and organisational barriers that so often prevent universities from achieving inclusiveness. One such barrier is the narrow understanding of disability and inclusive design. This research has contributed towards the development of a better understanding of these concepts in terms of individuals' differing abilities and needs in order to ensure that inclusiveness becomes standard practice in educational settings.

Rather than defining disability as a problem that an individual might have regarding how he/she can adapt to an existing built environment, this research has acknowledged disability as a matter of human difference. The interpretation of disability is thus centred on identifying the physical and mental barriers that prevent the person from gaining access to environments and services instead of focusing on his/her specific impairments. By acknowledging that human difference exists in terms of both physical and mental abilities, this research has pointed to the need for reconstruction of university settings so that they promote equality, human dignity and democracy by offering users flexibility and choice when it comes to using their facilities and services.

The current understanding of inclusive design focuses on eliminating physical barriers for individuals with disabilities and the elderly by making provisions that pay attention to their disabilities or needs. In exploring how to achieve an inclusive university, this study recognises the importance of going beyond the removal of architectural barriers for individuals with disabilities. Inclusive design, as the research has defined, is a holistic approach that roots itself in the university's built environment, culture, policy and practice, training, educational curriculum, communication and marketing strategy. A truly inclusive university is one that ensures inclusive access to its built environment by providing inclusive means of transportation, in particular bus stops and car parking bays that can accommodate a wide range of users. It also provides inclusive exterior routes, staircases and level access to the interior of its buildings, which have adjustable furniture, environmentally friendly material finishes, good ventilation systems, good-quality acoustics, adequate lighting, clearly signed emergency exit routes and an inclusive signage system. In addition, such a university provides inclusive educational information in a variety of text formats, including large print, Braille, audio tape, easy-to-read materials and electronic files. Inclusiveness is also reflected in the university training courses provided for its staff members that address disability awareness, legislation, sign language, lip reading and other users' needs. Generating a truly inclusive university involves detailed study of population distribution and trends, diverse lifestyles, needs and activities of users, and their interaction with the physical and educational environment. It also necessitates involving potential users in decision-making during the design and implementation phases.

8.4.3 Understanding Students' and Staff Members' Needs and Perceptions of Inclusive University Environment

Before this study was conducted, the population at the University of Kent had not been interviewed or surveyed to investigate their perceptions of and levels of satisfaction with the built environment. This study is the first to acknowledge such experiences and to highlight the physical, management and attitudinal barriers at the University of Kent. Obtaining such valuable information and feedback has been used as the basis for the proposed inclusive design strategies used in the case studies illustrated in Chap. 6. Moreover, the users' feedback and comments were used in drafting the inclusive design criteria presented in Chap. 7.

8.4.4   The Impact of Building Regulation on Achieving Inclusive Design

This study is the first to examine the impact of legislation and Building Regulation Approved Document M and their effects on removing architectural barriers for individuals with disabilities at British university buildings founded in the early 1960s. Six buildings from different decades, 1960s, 1970s, 1980s, 1990s, 2000s and 2010s, were studied to investigate whether the legislation and Building Regulation had managed to eliminate barriers for a wide range of users, including those with disabilities, and had thereby achieved inclusive design.

This study revealed that Approved Document M has influenced building designs which now incorporate new design features that were not included before the introduction of the regulation. Accessible parking bays, level access, automatic doors, reception desks with low and high counters, and accessible toilets are all new features that have been introduced to existing and new university buildings, but many of these features do not comply with the particular design specifications and measurements stipulated by Approved Document M. This study has highlighted that this Building Regulation has failed to completely achieve inclusive design in 1960s' university buildings, although it has led to accessibility enhancements for people with mobility and visual impairments.

To achieve inclusive design in a university setting, the research suggests that there is a need to revise the building regulations in the UK to extend the parameters of normal provision by taking account of the needs of people of different sexes, disabilities, religions, ages and family structures. The revised regulation should be applied to all building types, regardless of the age of the building and its construction condition. A building control team is required to review and check the accessibility proposals and their compliance with the proposed inclusive regulations. In addition, there should be a follow-through implementation scheme to monitor and check design compliance with inclusive design criteria.

8.4.5   Mainstreaming Inclusive Design in Educational Settings

This study shows how a fully inclusive university setting may be achieved, one that not only enhances the physical environment, but also offers learning environments that accommodate the needs of all users. Provision of modified keyboards, speech enhancement instruments, text-to-speech and notes in large print, in addition to Ipad tablets, software and other 'high-tech' gadgets and devices, can enable a diverse range of students with different abilities to communicate with their lecturers and gain access to learning and information.

To achieve inclusion in a university environment, the study proposes the establishment of an inclusive centre at the University of Kent. This centre would not only diagnose students with learning difficulties, but also provide inclusive design technologies to facilitate the access to learning for a broad range of users. Moreover, the centre would contribute to raising awareness of different disability needs among lecturers and staff members by providing training courses on enhancing the learning environment for all students, whatever their abilities. In addition, the study has contributed towards shifting attitudes towards designing inclusively, and to achieve that it proposed that inclusive design principles should become an integral part of the university's curriculum.

As for mainstreaming inclusive design in an educational setting, it is proposed that inclusive design principles should become part of the ranking criteria for universities. Those university institutions adopting these inclusive design criteria in their built environment, policies, practices and services would be allotted more points, thereby improving their ranking.

To conclude, a new holistic approach involving people with different abilities and of different ages, and collaborating with them, can result in new, inspiring and innovative environments that all can relish and enjoy. By adopting the inclusive design criteria that place users at the heart of the design process; that respond to their diversity and difference; that offer dignity, autonomy and choice; and that provide for flexibility of use, the ideal scenario of an inclusive university can be achieved:

> A place where I will be proud and really happy to say it doesn't matter what aid I am using, whether crutches, manual wheelchair, with a companion or an electric wheelchair, canes or guide dogs, the university is fully accessible wherever you go and whenever you want to.

Reference

Steinfeld, E., & Maisel, J. (2012). *Universal design: Creating inclusive environments*. Hoboken, NJ: John Wiley & Sons.

Appendix A: Jarman Building Access Statement

Hawkins\ Brown

13.0 Inclusive Access

As well as providing much needed facilities for the School of Drama, Film and Visual Arts the development will create a new 'Town Square' point of arrival and a new pedestrian priority spaces for the campus. This is an overview of the access issues relevant to the proposed building design and management.

Philosophy and approach to inclusive design – Statement of Intent

The proposal aims to meet the highest standards of accessibility and inclusion so that all potential users, regardless of disability, age or gender can use them safely and easily.

The proposal aims to promote inclusive access. Inclusive access is achieved by eliminating barriers, physical, attitudinal and procedural, which may otherwise inhibit the involvement of the whole community, not just disabled people. Inclusive access is about pro-actively reaching out to involve and include groups and individuals, who may feel that what you offer is currently 'not for people like them'. It means that organisations will need to consider their approach to all areas of their operations, including employment, policies, buildings and equipment, programming and marketing, and student development.

The ultimate aim of inclusive access is that the design and layout of the building should enable everybody to be able to enter it, use its facilities and leave safely, independently and with ease.

Relevant Legislation

The legislation that is of particular relevance is:

- Disability Discrimination Act (1995)
- Disability Discrimination Act (2005)
- The Building Regulations, Approved Document M, Access to and use of Buildings
- Town and Country Planning Act 1990

Other Sources of Guidance

- Designing for Accessibility (2004), Published by Centre for Accessible Environments, London.
- BS5588: Part 8:1988, Fire Precautions in the design, construction and use of buildings – Code of Practice for means of escape for disabled people, BSI, 1988
- BS8300: 2001, Design of Buildings and their approaches to meet the needs of disabled people. Code of Practice

Disability Discrimination Act (DDA) 1995

This Act makes '... it unlawful to discriminate against disabled persons in connection with employment, the provision of goods, facilities and services or the disposal or management of premises; to make provision about the employment of disabled persons.'

Disability Discrimination Act 2005

In addition to the DDA 1995, amendments in 2005 were designed to extend rights and provisions for disabled people. The Act also clarifies the various roles and responsibilities, as well as further clarifying the definition of 'disability'.

© Springer Nature Switzerland AG 2020
I. Shuayb, *Inclusive University Built Environments*,
https://doi.org/10.1007/978-3-030-35861-7

Hawkins\
Brown

Site, Approach, and Public Transport

Located within the heart of the University Campus the new School of Drama, Film and Visual will be well served by public transport. The building sits as part of a new pedestrianised 'town square'. University Road to the west of the site provides the principle traffic access to the campus. New provision will be created on University Road for drop off / collection immediately adjacent to the new building.

Arriving

The ultimate aim of inclusive physical access is that the design and layout of premises should enable everybody to be able to enter a building, use the facilities and leave safely, independently and with ease. The approaches to a building are of equal importance. Given that an estimated 73% of the disabled population use a private car as the most frequent mode of transport we have provide parking facilities close to the building. Drop off zones will also be provided near to the main entrance. Two disabled parking spaces are provided within to the rear of the building.

These parking bays will be:
- reserved for disabled people
- at least 3,600mm wide x ideally 6,000mm long and clearly marked out with access symbols and zone markings.
- monitored and controlled to prevent abuse
- set near dropped kerbs or with level access where necessary.

The drop off points to main entrance is also important. These spaces will:
- have an accessible route from the drop-off point to the building entrance free of obstacles
- clearly sign post the direction to the entrance
- provide seating along the route, if there are long travel distances between the drop-off point and entrance
- have pathways which provide a safe and obvious route to the building
- provide changes in paving at changes of direction, and tactile paving at dropped kerbs

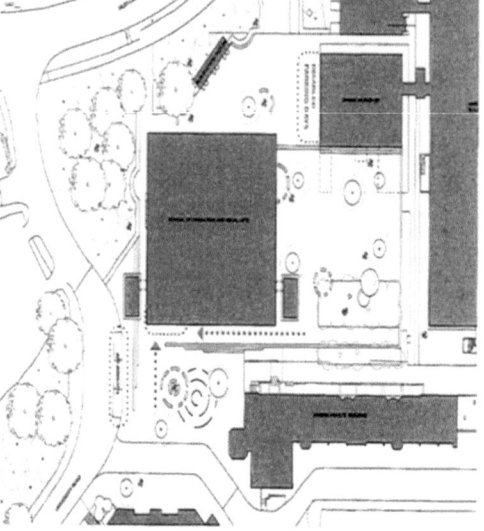

©Hawkins\Brown   February 2007

97

Entrances

The main entrance to the building will be designed to be accessible to everyone. We propose the use of automatic glass doors (not revolving as these are extremely hazardous), as these will assist most disabled people, including wheelchair users. The automatic doors should remain open for sufficient time to enable a blind person, a person with slow mobility, or a slow-moving wheelchair to pass through. In addition the main entrance doors will need to:

- be at least 900mm wide (between door stops)
- have a level landing outside
- have a level threshold
- have door controls and handles that are easy to see and at a height which can be reached by wheelchair users (no higher than 1,000mm from floor level)
- have at least 300mm alongside the leading edge of all doors to enable wheelchair users to open the door
- have a warning strip or logos at eye level for safety. Full glass doors and full height, large areas of glazing, can present particular access barriers for some disabled people. We aim to provide logos or safety markings at two heights: eye level (approximately 1,500mm from floor level for adults) and child/wheelchair user (approximately 1,200mm to make them visible)

Where any external entrances are locked or unattended it is important to ensure access for all. In these situations an intercom will be provided. We propose that intercoms are:

- at wheelchair accessible height (between 750mm and 1,000mm from floor level)
- have a solution to provide access for deaf people (links to a CCTV, minicom or video are useful for deaf visitors)

The entrance lobby has been designed to allow:

- a wheelchair user to clear the outer door before opening the inner door

We also highlight the importance of the lighting to the entrance lobby. The lighting in the lobby will need to be sufficient to help people adjust to changes in light between the outdoors and indoors.

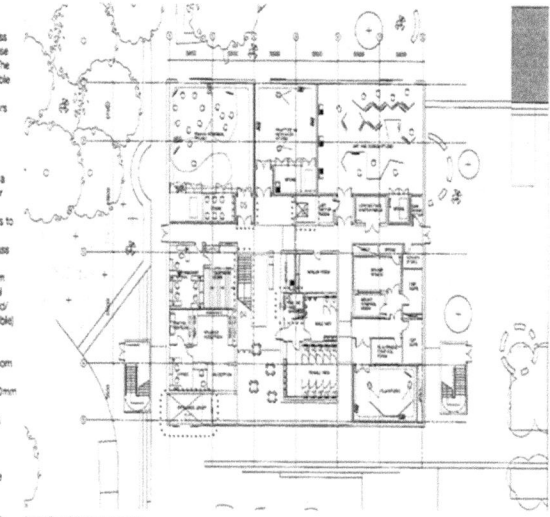

Ground Floor Plan showing key areas

01 Main Entrance Lobby - automatic sliding doors and level return

02 Level threshold to open access to piant needed via internal park ramp only

03 Accessible WC and Shower - located and set in accordance with Part M of the Building Regulations

04 Main circulation stair - designed to enclosed loaded use in accordance with Part M of the Building Regulations

05 Internal lobbies - sized in accordance with Part M of the Building Regulations

Hawkins\Brown   February 2007

58

Hawkins\ Brown

Lifts

The lift has been sized for independent use by a wheelchair user, enabling the user to enter the lift in a forward position and exit by reversing from the lift. The size of lift has been designed to reflect the capacity of the building.

Circulation

The circulation areas have been designed to ensure unobstructed access. The design considerations that have been taken into account are:

- all corridors have been designed to accommodate a minimum width of 1,200mm with no obstructions such as furniture or fire extinguishers.
- We propose that all Doors will:
  - have a minimum width of 900mm (between door stops) when fully open.
  - be fitted with vision panels to enable people to see and be seen.
  - will be fitted with lever type handles or 'D' pull handles at a height appropriate for a wheelchair user (1,000mm from floor level).
  - be light enough to be used by disabled people with limited mobility or strength.
- All internal lobbies are in accordance with Part M of the Building regulations

Stairs

Stairs will need to:

- be slip-resistant
- have a tactile surface to indicate the beginning and end of the flight
- be well lit.
- have the nosing strip of each step in a contrasting tone / colour to the tread (and ideally the risers should be of a different colour to the treads)

Handrails for stairs will need to:

- be at a height of 900mm (1,000mm at landings) on both sides running the entire length to enable those with a weakness on one side to use them.

Toilets and Showers

Each of these spaces have been designed to BS8300: 2001 and the requirements of Part M3 of the Building Regulations. A fully accessible toilet / shower needs to be designed to address the requirements of people with a variety of impairments. These spaces will be fully equipped for use by disabled people in both student and staff areas.

The standard dimensions used are:

- Unisex accessible corner WC layout 2200mm x 1500mm min.
- Accessible WC compartment for ambulant disabled people 800mm x 1500mm
- Self-contained shower room for independent use 2200mm x 2000mm

Signage and Navigation

We have developed the planning of the building to be simple and intuitive as the ability to navigate independently around a building is dependent upon the basic building layout.

We propose that any signage should be grounded in the following:

- the content of signs and information is written concisely and in plain English
- rules for clear print are followed (contrast between text and background colours, large enough text and easy-to-read fonts)
- simple illustrations or pictograms and symbols should be incorporated whenever possible
- all directional signs to and within the premises incorporate directional arrows
- signs are well lit with their own source of light

BS 8300: 2001 indicates that universally recognised symbols should be used to replace text, as an essential aid for people with learning difficulties. Where other types of pictograms and symbols are used these should be supplemented by text, and not used in isolation. The BS provides some examples. Further information on public information symbols can be found in BS 6034 and the RNIB publication Building Sight.

Lighting and Décor

Lighting and décor is important for navigation. Visually impaired people rely on being able to distinguish between the walls, floors, ceilings and doors, and between backgrounds and furniture.

Specifically:

- we propose the use of glare control measures such as blinds, matt finishes to combat reflection
- we aim to consider using colour as a means of assisting orientation, for instance, using a single colour for the floor surface to denote areas of circulation
- we aim to consider using changes of floor finish in a similar way as colour
- we aim to provide adequate contrast between doors, walls, floors and ceilings, and between furniture and the background against which it will be viewed
- we aim to develop a strategy to distinguish between trims such as coving, skirting boards, architrave, dado and handrails, door handles, finger and kick plates by use of colour, tonal and textural contrast

Appendix B: A Questionnaire Concerning the Accessibility of the Built Environment at the University of Kent, Canterbury Campus

This questionnaire has been designed to find out the views of disabled and non-disabled individuals and their experiences on accessibility at the University of Kent, Canterbury campus.

The research aims are twofold: (1) to investigate experiences of disabled individuals who study and work at the University of Kent, Canterbury campus, and (2) to help the researchers and designers improve the accessibility of the University built environment.

We do encourage all individuals who have come across barriers at the university campus because of at least one of the below-mentioned disabilities or barriers to take part in the current questionnaire:

1. Mobility difficulties/wheelchair users
2. Visual impairments
3. Hearing impairments
4. Learning difficulties
5. Multiple disabilities (visual, hearing, mobility problem, and/or cognitive learning disability)
6. Pregnant woman
7. Mother with pushchair

Personal Information

1. Are you

☐ Female
☐ Male
☐ Do not wish to declare

© Springer Nature Switzerland AG 2020
I. Shuayb, *Inclusive University Built Environments*,
https://doi.org/10.1007/978-3-030-35861-7

2. Please tick your age group category:

☐ Under 18   ☐ 35–49
☐ 18–24   ☐ 50–64
☐ 25–34   ☐ 65 and above

3. Are you

☐ British
☐ Non-British

4. Do you consider yourself to have a disability?

☐ Yes
☐ No
☐ Partly

5. If yes please tick all those applying to you:

☐ Mobility difficulties/wheelchair users
☐ Visual impairment
☐ Hearing impairment
☐ Multiple disabilities (please tick if you have at least two of the following disabilities: visual, hearing, mobility difficulties/wheelchair users, specific learning disability)
☐ Specific learning disability/e.g.: (dyslexia)
☐ Mental health difficulty
☐ You have a disability that cannot be seen (e.g.: diabetes, epilepsy, heart condition)
☐ You have a disability, special need or medical condition that is not listed
☐ Pregnant woman
☐ Mother with pushchair

If you would prefer to describe your impairment, condition or disability in your own words, please do so here.

6. Did your disability start

☐ From birth
☐ After birth before joining university
☐ After birth after joining university
☐ Not applicable

7. Do you have any additional support needs to enable you to study/work or to take exams? Please give details.

8. Are you

☐ Undergraduate student
☐ Postgraduate student
☐ Member of staff

☐ Mother/carer
☐ Visitor
☐ Visitor

9. Please rate your knowledge of the rights that are endorsed in the British legislation (Equality Act 2010).

☐ Very knowledgeable
☐ Knowledgeable
☐ Somewhat knowledgeable
☐ Not knowledgeable

10. If you are very knowledgeable/knowledgeable with the rights of the Equality Act (2010) please rate your satisfaction with the Act.

☐ Very satisfied
☐ Fairly satisfied
☐ Satisfied
☐ Not satisfied

Assessing the University Buildings and Signage

11. Do you think that you have the same chances in accessing the Canterbury campus as anyone else?

☐ Yes
☐ No

12. Do you feel that the built environment and buildings at the University of Kent, Canterbury campus, take into account the needs of individuals with disabilities?

☐ Yes
☐ No

13. Please tell us about your experience of accessing the university campus and buildings.
    Think about two buildings you would use most often.

Building 1	Building 2
(Name of building)	(Name of building)

Please rate your experience in accessing this building

Building 1	Building 2
☐ Excellent	☐ Excellent
☐ Very good	☐ Very good
☐ Good	☐ Good
☐ Poor	☐ Poor
☐ Very poor	☐ Very poor

Please explain why

14.   How would you usually travel to the university campus? (Please tick only one.)

☐ Mainly by car
☐ Mainly by public transport
☐ Other (e.g. by foot or bicycle) Specify_____

Answer Qs a–c If You Travel to the University Campus by Car. If Not, Move to Q 20

(a) Please rate your experience in accessing the car parking at the university campus.

☐ Excellent
☐ Very good
☐ Good
☐ Poor
☐ Very poor

Please explain why

(b)   Is the position of the designated accessible car parking bays sited as close to the main entrance of the university?

☐ Yes
☐ No

(c) Are the ticket machines fully accessible to disabled people?

☐ Yes   ☐ No

15. Thinking about your experience, please rank the following barriers to indicate what you think is the most disabling barrier that hinders you from accessing the university buildings.

1 = Top disabling barrier
2 = Your second
3 = Your third
4 = Your fourth
5 = Your fifth
6 = Your sixth
7 = Your seventh
8 = Your eighth
9 = Your ninth

- ☐ Stairs
- ☐ Steep ramps
- ☐ Kerbs
- ☐ Handrails
- ☐ Signs
- ☐ Doors
- ☐ WC
- ☐ Changing room for babies
- ☐ Not providing induction loops

16. The research aims to ensure that there is adequate and appropriate signage to and within the buildings. Please indicate which of the following best describes your experience of this (please tick only one).

   - ☐ Excellent
   - ☐ Very good
   - ☐ Good
   - ☐ Poor
   - ☐ Very poor

17. Are you aware of the emergency egress/exit location which operates in the building(s) you use?

   - ☐ Yes
   - ☐ No

18. Are the signs which mark emergency routes and exits clear enough?

   - ☐ Yes
   - ☐ No

19. In your opinion, which of the below two design approaches would you recommend in the design progress?

   - ☐ Accessible design for only disabled individuals
   - ☐ Inclusive design for all (disabled and non-disabled individuals)

20. In your opinion do you favour a design feature that:

   - ☐ Incorporates seamlessly with the buildings
   - ☐ Is specially marked to incorporate your special needs

21. In your opinion do you think that the design features adopted at university buildings help in community integration between individuals with and without disability?

   - ☐ Yes
   - ☐ No

22. Which of the following buildings you think the researcher should prioritise to start with the access improvements. Please tick the relevant boxes in the list below which you consider as the most important to you.

Type of building	Name of building	Very important	Important	Not bothered	Do not use	Would use if accessible
Leisure centre	Students Union	■ ☐	■ ☐	■ ☐	■ ☐	■ ☐
Registry offices	Registry	■ ☐	■ ☐	■ ☐	■ ☐	■ ☐
Colleges	Rutherford College Eliot College	■ ☐ ■ ☐	■ ☐ ■ ☐	■ ☐ ■ ☐	■ ☐ ■ ☐	■ ☐ ■ ☐

Appendices
Appendix B
A questionnaire concerning the accessibility of the built environment at the University of Kent, Canterbury Campus

Library	Templeman Library	■ ☐	■ ☐	■ ☐	■ ☐	■ ☐
Theatre	Gulbenkian Theatre	■ ☐	■ ☐	■ ☐	■ ☐	■ ☐
Department	Marlowe Building	■ ☐	■ ☐	■ ☐	■ ☐	■ ☐
New buildings	The Arts Building	■ ☐	■ ☐	■ ☐	■ ☐	■ ☐

23. There are many things the researcher can provide to make the buildings accessible; below is a list of some of those; please tick each item to show how important they are to you.

Item	Important	Would be helpful	Not necessary
Level access	■ ☐	■ ☐	■ ☐
Braille and tactile signs	■ ☐	■ ☐	■ ☐
Handrails and tap rails	■ ☐	■ ☐	■ ☐
Automatic doors	■ ☐	■ ☐	■ ☐
Dropped kerbs	■ ☐	■ ☐	■ ☐
Colour contrast	■ ☐	■ ☐	■ ☐
Permanent ramps	■ ☐	■ ☐	■ ☐
Temporary ramps	■ ☐	■ ☐	■ ☐

Appendices
Appendix B
A questionnaire concerning the accessibility of the built environment at the University of Kent, Canterbury Campus

Wheelchair accessible	■ ☐	■ ☐	■ ☐
Lifts	■ ☐	■ ☐	■ ☐

Item	Important		Would be helpful		Not necessary	
Good acoustics	▧	☐	▧	☐	▧	☐
Low-level counters	▧	☐	▧	☐	▧	☐
Induction loops	▧	☐	▧	☐	▧	☐
Intercom systems	▧	☐	▧	☐	▧	☐
Good lighting	▧	☐	▧	☐	▧	☐
Signs with both pictures and words	▧	☐	▧	☐	▧	☐
Blue Badge parking	▧	☐	▧	☐	▧	☐

Appendices
Appendix B
A questionnaire concerning the accessibility of the built environment at the University of Kent, Canterbury Campus

Item	Important		Would be helpful		Not necessary	
Low-level switches	▧	☐	▧	☐	▧	☐
Texture contrasts	▧	☐	▧	☐	▧	☐
Signs with words	▧	☐	▧	☐	▧	☐
Steps with corduroy tactile paving	▧	☐	▧	☐	▧	☐
Signs with pictures	▧	☐	▧	☐	▧	☐

24. These are just some suggestions. Do you have any others that you would like the researcher to consider? If so, what are they?

Contact Details

This information is required to enable us to feedback the results of this questionnaire and/or include you in further survey methods *in the form of an access audit* or consultations.

The purpose of an access audit is to establish how well a particular building or environment performs in terms of access and ease of use by a wide range of potential users including individuals with disabilities in order to recommend access improvements.

This contact information will form a confidential contact list only separate from your questionnaire answers above.

First Name:
Last Name:
Address:
Home Telephone Number:
Work Telephone Number:
Mobile Telephone Number:
Fax Number:
E-mail Address:
How would you prefer us to contact you about being involved in this work in the future?

Thank you for helping us with this questionnaire

Appendix C: Interview Questions for Individuals with Disabilities at the University of Kent

Personal Information and Background

1. Are you an undergraduate/postgraduate/staff?
2. What is your course of study? Or your profession?
3. When did your disability start?
4. How long have you been enrolled in this university?

Evaluation and Experience in Accessing the University

5. Have you come across a physical barrier at the University? If yes, can you briefly describe your experience?
6. What is the first barrier that you come across when accessing the University?
7. What are the main barriers you face when accessing the entrance of the buildings you use most (entry systems, such as induction loops, swipe card, LED or text display; entrance doors, width, pressure needed to open doors; doors if they are glass, automatic or revolving; thresholds flush or not, doormats, visual contrast)?
8. What are the main barriers you face when accessing the reception area of the buildings you use most: accessibility of lobby and exits; reception area seating and lighting; reception desks (lowered section, adequate knee recess, induction loops)?
9. What are the main barriers you face when accessing the corridors and passageways of the building you most use (corridor width, doors open into corridor or not, corridors are splayed or rounded at corners, surface finishes if slip resistant, firm, refectory, visual contrast between wall and floor, handrails along the length of corridors, fire doors whether can be opened easily, lighting in corridors, signage)?
10. What are the main barriers you face using the stairs and steps in the building you most use (if stairs are spiral, or contain a section which has winders, stairs have open risers, if stairs are narrow in their width, corduroy tactile flooring, change in floor texture on top and bottom of each flight, handrails oval, circular or rectangular, nosings on each step are adequately visually contrasted)?

© Springer Nature Switzerland AG 2020

I. Shuayb, *Inclusive University Built Environments*,

https://doi.org/10.1007/978-3-030-35861-7

11. What are the main barriers you face when accessing internal and external ramps (gradient of ramp, if it is temporary or permanent, surface finishes if different, tactile warning being provided on top or bottom of ramp, handrails provided, handrail shape if oval, circular, or rectangular, steps provided as alternative to using ramp, lighting, visual contrast)?

12. What are the main barriers you face when using lifts at buildings you most use (manoeuvring, floor surface of lift firm or slip resistant, handrail provided, position of buttons, audible sound, mirror, visual contrast and lighting, emergency equipment, text phone, induction coupler)?

13. What are the main barriers you face when accessing the toilets at the building you most use (signage indicating the route, visual contrast, wall tiles or finishes not reflective, lever mixer taps, floor finish is slip resistant, flushes to WC and light switches, lighting, height of basin, height of toilet seat)?

14. What are the main barriers you face when accessing the wheelchair-accessible toilet at the building you most use (doors open outwards, or indoors, height of seat, flush type, grab rail, lighting, coat hook, general spatial layout, height of basin, soap dispenser, single-lever mixer tap provided, paper towel dispenser can be reached whilst seated, lighting, white pull cord or automatically device switches on light, alarm call system, visual contrast, floor surface slip resistant, small shelf, mirror)?

15. What are the main barriers you face when accessing the **Templeman Library**?

16. What are the main barriers you face when accessing the **Eliot College**?

17. What are the main barriers you face when accessing the **Venue Student Union**?

18. What are the main barriers you face when accessing the **Registry**?

19. What are the main barriers you face when accessing Jarman **Building**?

20. What are the main barriers you face when accessing the Marlowe **Building**?

21. **The New Arts Jarman building** is a newly built building: Do you think that it incorporates accessible design if compared to the older buildings at the University?

Interaction Experiences with Staff Members

22. What are your views on staff members, such as lecturers and receptionists? Are they helpful, knowledgeable, friendly and polite, when dealing with you?

23. Have you faced any kind of direct discrimination from students/staff members? If yes, did you report about such discrimination to student affairs office? If yes what kind of action did the student affairs office take?

24. How could you help educate/raise awareness for non-disabled people?

Means of Escape

25. Have you submitted a means of escape personal plan when you joined the university?

26. Are you aware of the emergency egress/exit location which operates in the building(s) you use? If yes do you think that the signs which mark emergency routes and exits are clear enough?

Design Approach

27. In your opinion how could the built environment at the **University of Kent, Canterbury campus**, incorporate accessible design?
28. In your opinion how can the university buildings create a friendly and welcoming environment that suits all its students?
29. In your opinion, what kind of design feature you suggest? Do you favour a design feature that incorporates seamlessly with the buildings, or do you favour a special design that incorporates your special need only?
30. Describe an ideal accessibility scenario.

Index

A
Access and facilities for disabled people, 37
Access audits, 205
   analysis, 75
   assessment, 63, 66
   building and services, 81
   buildings, selection, 69
   categories and subcategories, 76
   description, 69
   disability classification, 85
   external physical environment
      bus drop-off zone, 86
      car parking, 86–88
      external ramps, 88, 89
      external steps, 88
   features, 81, 83, 84
   inclusive design principles, 70, 84
   internal physical environment
      access statement for the Jarman
         Building, 96
      arriving, 97
      corridors, 90, 91
      emergency egress routes, 96
      entrance, 98–99
      internal staircases, 92, 93
      main entrances, 89, 90
      passenger and platform lifts, 92, 93
      reception counter, 90, 91
      site approach and public transport, 96
      toilet compartments, female and male,
         94
      unisex wheelchair-accessible toilets,
         95, 96
   journey cycle, 71
   management practices and procedures, 99,
      100
   physical environment components, 83
   physical features and management
      procedures, 70
   rating system, 72
   role of auditor, 70
Accessibility
   barriers, 188
   Building Regulations, 187
   in campus, 184
   questionnaire, 199
   shared task, 188
Accessibility barriers
   access audits (*see* Access audits)
   external physical environment
      buses, 103
      external ramps, 103, 104
      external steps, 103
      parking spaces, 102, 103
   internal barriers
      corridors, 106
      counters, 106
      emergency egress, 112
      female and male compartments, 110,
         111
      internal staircases and lifts, 106
      lifts, 108, 109
      main entrance, 104
      staircases, 108
      toilet compartments, 109
      wheelchair-accessible toilets, 111
   Part M standards, 88
   stakeholder consultations (*see* Stakeholder
      consultations)
   students with disabilities (*see* Students and
      staff members with disabilities)
   unisex wheelchair-accessible toilets, 95

© Springer Nature Switzerland AG 2020
I. Shuayb, *Inclusive University Built Environments*,
https://doi.org/10.1007/978-3-030-35861-7